77 Advances in Polymer Science
Fortschritte der Hochpolymeren-Forschung

Induced Circular Dichroism in Biopolymer-Dye Systems

By M. Hatano

Editor: S. Okamura

With 73 Figures and 11 Tables

Springer-Verlag
Berlin Heidelberg New York Tokyo

Prof. Dr. Masahiro Hatano
The Chemical Research Institute of
Non-Aqueous Solutions
Tohoku University
Sendai 980/Japan

ISBN-3-540-15854-5 Springer-Verlag Berlin Heidelberg New York Tokyo
ISBN-0-387-15854-5 Springer-Verlag New York Heidelberg Berlin Tokyo

Library of Congress Catalog Card Number 61-642

This work is subject to copyright. All rights are reserved, whether the whole or part of the material is concerned, specifically those of translation, reprinting, re-use of illustrations, broadcasting, reproduction by photocopying machine or similar means, and storage in data banks. Under § 54 of the German Copyright Law where copies are made for other than private use, a fee is payable to "Verwertungsgesellschaft Wort", Munich.

© by Springer-Verlag Berlin Heidelberg 1986
Printed in GDR

The use of registered names, trademarks, etc. in this publication does not imply, even in the absence of a specific statement, that such names are exempt from the relevant protective laws and regulations and therefore free for general use.

Typesetting and Offsetprinting: Th. Müntzer, GDR
Bookbinding: Lüderitz & Bauer, Berlin
2154/3020-543210

Editors

Prof. Henri Benoit, CNRS, Centre de Recherches sur les Macromolecules, 6, rue Boussingault, 67083 Strasbourg Cedex, France

Prof. Hans-Joachim Cantow, Institut für Makromolekulare Chemie der Universität, Stefan-Meier-Str. 31, 7800 Freiburg i. Br., FRG

Prof. Gino Dall'Asta, Via Pusiano 30, 20137 Milano, Italy

Prof. Karel Dušek, Institute of Macromolecular Chemistry, Czechoslovak Academy of Sciences, 16206 Prague 616, ČSSR

Prof. John D. Ferry, Department of Chemistry, The University of Wisconsin, Madison, Wisconsin 53706, U.S.A.

Prof. Hiroshi Fujita, Department of Macromolecular Science, Osaka University, Toyonaka, Osaka, Japan

Prof. Manfred Gordon, Department of Pure Mathematics and Mathematical Statistics, University of Cambridge CB2 1SB, England

Prof. Gisela Henrici-Olivé, Chemical Department. University of California, San Diego, La Jolla. CA 92037, U.S.A.

Prof. Dr. habil. G. Heublein. Sektion Chemie. Friedrich-Schiller-Universität. Humboldtstraße 10, 69 Jena, DDR

Prof. Dr. Hartwig Höcker, Lehrstuhl für Textilchemie und Makromolekulare Chemie, RWTH, Worringer Weg 1, 5100 Aachen, FRG

Prof. Hans-Henning-Kausch, Laboratoire de Polymères, Ecole Polytechnique Fédérale de Lausanne. 32. ch de Bellerive, 1007 Lausanne, CH

Prof. Joseph P. Kennedy, Institute of Polymer Science, The University of Akron, Akron, Ohio 44325, U.S.A.

Prof. Anthony Ledwith, Department of Inorganic, Physical and Industrial Chemistry. University of Liverpool, Liverpool L69 3BX, England

Prof. Seizo Okamura, No. 24, Minamigoshi-Machi Okazaki, Sayko-Ku. Kyoto 606, Japan

Prof. Salvador Olivé, Chemical Department, University of California, San Diego, La Jolla, CA 92037, U.S.A.

Prof. Charles G. Overberger, Department of Chemistry, The University of Michigan, Ann Arbor, Michigan 48104, U.S.A.

Prof. Helmut Ringsdorf, Institut für Organische Chemie, Johannes-Gutenberg-Universität. J.-J.-Becher Weg 18–20. 6500 Mainz, FRG

Prof. Takeo Saegusa, Department of Synthetic Chemistry, Faculty of Engineering, Kyoto University, Kyoto, Japan

Prof. Günter Victor Schulz, Institut für Physikalische Chemie der Universität, 6500 Mainz, BRD

Prof. William P. Slichter, Chemical Physics Research Department, Bell Telephone Laboratories, Murray Hill, New Jersey 07971, U.S.A.

Prof. John K. Stille, Department of Chemistry, Colorado State University, Fort Collins, Colorado 80523, U.S.A.

Editorial

With the publication of Vol. 51, the editors and the publisher would like to take this opportunity to thank authors and readers for their collaboration and their efforts to meet the scientific requirements of this series. We appreciate the concern of our authors for the progress of "Advances in Polymer Science" and we also welcome the advice and critical comments of our readers.

With the publication of Vol. 51 we would also like to refer to a editorial policy: *this series publishes invited, critical review articles of new developments in all areas of Polymer Science in English (authors may naturally also include works of their own)*. The responsible editor, that means the editor who has invited the author, discusses the scope of the review with the author on the basis of a tentative outline which the author is asked to provide. The author and editor are responsible for the scientific quality of the contribution.

Manuscripts must be submitted in content, language, and form satisfactory to Springer-Verlag. Figures and formulas should be reproducible. To meet the convenience of our readers, the publisher will include "volume index" which characterizes the content of the volume.

The editors and the publisher will make all efforts to publish the manuscripts as rapidly as possible, i. e., at the maximum six months after the submission of an accepted paper. Contributions from diverse areas of polymer science must occasionally be united in one volume. In such cases a "volume index" cannot meet all expectations, but will nevertheless provide more information than a mere volume number.

Starting with Vol. 51, each volume will contain a subject index.

Editors   Publisher

Preface

The interaction of biopolymers with dyes has been investigated by many researchers from the viewpoints of physiology, biochemistry, and physical chemistry. These investigations have been made mainly by spectroscopic techniques, since the interactions can be studied in detail even if the concentration of dyes is very low. The electronic spectra of the biopolymer-dye systems very often differ from those of free dyes in a number of interesting ways. In the biopolymer-dye systems, the dyes often lose absorption intensity (hypochromism), change the absorption maxima (methachromasy), display new bands, lead to fluorescence quenching, and under certain conditions become *optically active*. The last property is, perhaps, the most significant, since the biopolymers are generally *chiral*. Though a symmetric chromophore having a plane or center of symmetry is optically inactive as it is, it becomes optically active when placed in an asymmetric field of other chiral molecule(s). This effect has been known as the *induced optical activity* or *induced circular dichroism* (ICD). In the biopolymer-dye systems, a number of planar dyes without any chirality become optically active.

The measurement of ICD is quite useful to analyze the interaction between a biopolymer and a given guest molecule. First, in this article theoretical considerations of ICD are discussed, followed by various examples of ICD phenomena for various kinds of biopolymer-achiral molecule systems which are classified into several groups. General aspects of the experiments and apparatus for circular dichroism (CD) are also treated.

Sendai, October 1985   M. Hatano

Table of Contents

1 **Introduction** 1

2 **Optical Activity** 2
   2.1 Optical Activity and Molecular Symmetry 2
   2.2 Polarized Light. 3
   2.3 Optical Rotation 4
   2.4 Circular Dichroism 7

3 **Theory of Optical Activity** 9
   3.1 Symmetry Considerations 10
   3.2 Coupled Oscillator Mechanism 12
   3.3 The Asymmetrically Perturbed Field Mechanism 21
   3.4 Induced Optical Activity. 22
      3.4.1 Dispersion Force-Induced Circular Dichroism (DICD) 24
      3.4.2 Hydrogen Bonding-Induced Circular Dichroism (HBICD). . . . 26
      3.4.3 Ionic Coupling-Induced Circular Dichroism (ICICD) 27
      3.4.4 Ligation-Induced Circular Dichroism (LICD) 27
      3.4.5 Charge-Transfer-Induced Circular Dichroism (CTICD) 29
      3.4.6 Liquid Crystal-Induced Circular Dichroism (LCICD) 34
      3.4.7 Hydrophobic Interaction-Induced Circular Dichroism (HIICD) 37

4 **Induced Circular Dichroism in Nucleic Acid-Dye Systems** 40
   4.1 Circular Dichroism Spectral Properties of Nucleic Acids 41
   4.2 Feasibility of the Induced Circular Dichroism Technique for Nucleic Acid
      Research 44
   4.3 Conformational Changes of Deoxydinucleoside Monophosphate and
      Polydeoxyribonucleotides Induced by Added Drugs or Metal Ions . . . 45
   4.4 Metal Complex Binding to DNA 47
   4.5 Nucleic Acid-Protein Systems 49
   4.6 Future Trends and Scope on Induced Circular Dichroism in Nucleic
      Acid-Dye Systems 49

5 **Induced Circular Dichroism in Protein-Dye Systems** 50
   5.1 Estimation of the Contents of Each Fraction and of the Secondary
      Structure 51
   5.2 Prediction of Secondary Structure in Proteins 54

    5.3 Conformational Change of Proteins 61
    5.4 Induced Circular Dichroism in Side-Chain Chromophores 66
    5.5 Induced Circular Dichroism of Aromatic Compounds Bound to Proteins . 72
    5.6 Induced Circular Dichroism of Heme and Chlorophyll Bound to Proteins . 78
    5.7 Metal Binding to Proteins 87
    5.8 Future Trends and Scope on Induced Circular Dichroism in
        Protein-Dye Systems 88

6 Induced Circular Dichroism in Polysaccharide-Dye Systems 89
    6.1 Vacuum Ultraviolet Circular Dichroism and Its Application to
        Saccharides 89
    6.2 Dye Probes for Polysaccharide Conformation Analysis 90
    6.3 CD Analysis of Side-Chain Chromophores on Saccharides 90
    6.4 Benzoates Exciton Rule for Determining the Anomeric Configuration in
        Saccharides 93
    6.5 Future Trends and Scope on Circular Dichroism in Saccharides . . . 93

7 Induced Circular Dichroism in Liquid Crystalline Phases 94
    7.1 The Relationship Between the CD Sign and the Helical
        Sense in Cholesteric Phases 94
    7.2 The Relationship Between the CD Signs and the Sign of the
        Pitch-Band CD 97
    7.3 Enhanced Circular Dichroism of Aggregates of Chiral Amphiphiles . . 99
    7.4 Future Trends and Scope on Liquid Crystal-Induced Circular Dichroism 100

**8 Experimental Considerations and Apparatus for Measuring
    Circular Dichroism** 102
    8.1 Experimental Considerations and Precautions 102
    8.2 Aparatuses 104

**9 Concluding Remarks and Future Trends on Induced
    Circular Dichroism** 106

**10 Appendix: Magnetic Circular Dichroism Techniques Coupled
    with the Circular Dichroism Techniques** 107

11 References 114

Author Index Volume 1—77 123

Subject Index 133

    Erratum 136

Abbrevations

ICD	induced circular dichroism,
CD	circular dichroism,
ORD	optical rotatory dispersion,
LCP	left-circularly polarized light,
RCP	right-circularly polarized light,
DICD	dispersion force-induced circular dichroism,
HBICD	hydrogen bonding-induced circular dichroism,
ICICD	ionic coupling-induced circular dichroism,
LICD	ligation-induced circular dichroism,
CTICD	charge-transfer-induced circular dichroism,
LCICD	liquid crystal-induced circular dichroism,
HIICD	hydrophobic interaction-induced circular dichroism,
MCD	magnetic circular dichroism,
NMR	nuclear magnetic resonance,
SOR	synchrotron orbital radiation,
MO	molecular orbital,
LUMO	lowest unoccupied molecular orbital,
CT	charge-transfer,
HPLC	high performance liquid chromatography,
TCNE	tetracyanoethylene,
P-450	cytochrome P-450,
β-CDx	β-cyclodextrin,
Tris	2-amino-2-hydroxymethyl-1,3-propanediol, tris(hydroxymethyl)aminomethane,
EDTA	N,N'-1,2-ethanediylbis[(N-carboxymethyl)glycine], (ethylenedinitrilo)tetraacetic acid, ethylenediaminetetraacetic acid, edetic acid, the ionic form of which is described as *edta*,
PBLG	poly(γ-benzyl-L-glutamate),
PBDG	poly(γ-benzyl-D-glutamate),
PLL	poly(L-lysine),
PLB	poly(α,γ-diamino butyric acid),
PCLG	poly{γ-[2-(9-carbazoyl)ethyl]-L-glutamate,
PLNA	poly(L-naphthylalanine),
PLGA	poly(L-glutamic acid),
AO	3,6-bis(dimethylamino)acridine, acridine orange,
MO	4-[4-dimethylamino)-phenylazo]benzenesulphonic acid,

PNVC	poly(*N*-vinylcarbazole),
EtCz	*N*-ethylcarbazole,
BChl	bacteriochlorophyll,
Chl	chlorophyll,
HFIP	hexafluoro-2-propanol,
ICM	intracytoplasmic membrane,
VUV	vaccum ultraviolet region,
CMC	critical micelle concentration.

1 Introduction

Most biologically important molecules are *optically active* or *chiral*. Such molecules are not identical to their mirror images. This *chirality* in the *optically active* molecules results from the presence of asymmetric atom(s) or molecular dissymmetry. For example, a carbon tetrahedrally bound to four different atoms or groups can exist in two different structures that are mirror images of one another. Each structure is called *"enantiomer"* of the given compound. Chirality also occurs in molecules or ions without asymmetric carbons. Thus, tris(ethylenediamine)cobalt(III) ions have two types of enantiomeric isomers of Δ and λ types, which are distinguished by *"handness"* of the ligands around the central metal ion. Similar handness exists in biopolymers such as polynucleotides or proteins. An important source of the handness in the biopolymers is whether the helical backbone linkage winds in a right- or left-handed sense. Most biopolymers wind in the right-handed sense. The origin of this handness in biopolymers lies in the fact that the chiral monomeric units, nucleotides, or α-amino acid residues, are all of one type; the D-enantiomers of sugars in nucleotides and the L-enantiomers of amino acids in proteins.

Molecules that are asymmetric or dissymmetric exhibit *optical rotation*, which depends on the difference in real refractive indeces of a given molecule for left- and right-circularly polarized light (LCP and RCP), and *circular dichroism* (CD), which is the difference in imaginary refractive indeces (absorption) of the two types of circularly polarized light, LCP and RCP. The wavelength dependence of optical rotation is known as *optical rotatory dispersion* (ORD). The handness in chiral molecules or ions can interact with the circularly polarized light, LCP or RCP, selectively in one sense. The chirality or handness of chiral molecules or ions in solution can be distinguished by a different interaction with the light that is polarized in a chiral way, left- or right-circularly polarized. Both CD and ORD are referred to as natural optical activity. In another way, this optical activity has been extensively found in systems of achiral molecules bound to a chiral molecule in solution.

It is well known that optical activity is induced in achiral species in the presence of chiral species, giving rise to *"induced circular dichroism* (ICD)" for the absorption band(s) of the achiral species. As early as 1965, Mason et al.[1] observed CD bands for the d-d transitions of $[Co(NH_3)_6](ClO_4)_3$ in aqueous diethyl-(+)-tartrate. They found this ICD to be due to an outer-sphere coordination. ICD spectra also have been investigated by Bosnich et al.[2,3] not only for the d-d transitions of $[PtCl_4]^{2-}$ but also for the n-π* transitions of benzil and benzophenone in (S,S)-2,3-butanediol. Extensive studies on ICD have been carried out by Hayward et al.[4] for the n-π* transitions of symmetric and racemic aliphatic ketones in chiral tetrahydrofuranols. Noack[5] investigated the concentration dependence of ICD bands, and he suggested that a type of molecular complex with a 1:1 molar ratio exists in the system of achiral saturated ketones and chiral *l*-menthol. These ICD experiments have so far been carried out for various chiral media. Axelrod et al.[6], Bolard[7,8], and Hayward et al.[9,10] have observed the ICD bands for the n-π* transitions of carbonyl and nitro compounds in such solvents as CCl_4 or CH_3CN with optically active substances. They concluded that intermolecular interactions such as hydrogen bonding, van der Waals interactions, or ionic coupling between chiral species and achiral ones can induce the optical activity for the electronic transitions of achiral species. Schipper and Nordén [11] quite recently have further discussed the ICD mechanism.

ICD phenomena have been found for many kinds of biopolymer-dye systems [12]. Natural organic substances are generally endowed with a chirality which is restricted to one sense when they are biosynthesized, and they become optically active. This is due to the fact that the chirality of the naturally occurring molecules is necessary to construct a suitable structure for controlling or fulfilling their functions in living tissues. Accordingly, information on the chirality or secondary structures of biomolecules is necessary to elucidate their physiological activities. This ICD technique can provide useful information on the chirality of biomolecules when an achiral molecule is added to the biomolecule with unknown chirality. Further, we can obtain information on the binding mode of a given molecule, which is mimicking a substrate, with the biomolecule through the observed ICD.

2 Optical Activity

2.1 Optical Activity and Molecular Symmetry

If a molecule is rotated around an axis and the resulting orientation is reflected in a plane perpendicular to this axis, and if the resulting orientation is superimposable on the original, the molecule is said to possess a *rotational-reflection axis*. This axis is designated as S_n, and is also called an *alternating* or *improper axis*, in which n is the order (n fold). The traditional definition of *optical activity* is to examine the superimposability of the original with its mirror image. The lack of this superimposability, which is required for *optical activity*, indicates the absence of an *alternating axis* of any order, S_n. Conversely, however, the existence of an S_n axis is insufficient to indicate *optical activity*. All molecules which lack a *rotational-reflection axis* of any order, S_n, are said to be *dissymmetric*, *chiral*, or *optically active*. Such a molecule and its mirror image cannot be superimposed sterically by any kind of rotational or translational operation. For example, tris(ethylenediamine)Co(III) ions have C_3 axes (neglecting the conformational change of ethylene units), but it does not possess any S_n axis, though these ions have no *asymmetric carbon*; the ions are dissymmetric or chiral, but not asymmetric (Fig. 1). The only symmetry that dissymmetric molecules possess is one or more C_n axes. For example, chiral biaryls and tris(ethylenediamine)-Co(III) have a C_2 axis and three C_3 axes together with three C_2 axes, respectively.

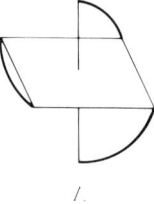
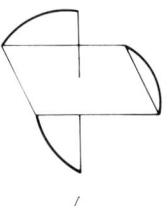

Fig. 1. The absolute configurations of two optical isomers of a tris-bidentate complex with symmetrical ligands

2.2 Polarized Light

The electric vector for a plane wave propagating in the positive z-direction can be represented as:

$$\vec{E}(t, z) = \text{Re } \vec{u} \, E_0 \exp[-i\omega(t - z/c)], \tag{1}$$

where Re means real parts in the large bracket on the right side of Eq. (1), $\vec{u}$ is a vector describing the polarization state of the wave, and E_0 is the amplitude of the wave. Equation (1) can be converted to Eq. (2):

$$\vec{E}(t, z) = \text{Re } \vec{u} \, E_0[\cos \omega(t - z/c) - i \sin \omega(t - z/c)]. \tag{2}$$

If the right-handed Cartesian frame is chosen (Fig. 2) and is assumed as

$$\vec{u} = (1/\sqrt{2}) \begin{pmatrix} 1 \\ -i \end{pmatrix}, \tag{3}$$

we obtain Eqs. (4) and (5):

$$E_x(t, z) = (E_0/\sqrt{2}) \cos \omega(t - z/c), \tag{4}$$

$$E_y(t, z) = (-E_0/\sqrt{2}) \sin \omega(t - z/c). \tag{5}$$

The propagation of the wave can be considered at a *fixed value of z*, say $z = 0$ for convenience. Equations (4) and (5) show that at $t = 0$ the electric vector lies along the x-axis, while at a later time $t = \pi/2\omega$ it is along the negative y-axis, and at a still later time $t = \pi/\omega$ it is along the negative x-axis, and so on. Thus, as viewed from a point on the positive z-axis, looking back towards the origin ($z = 0$), the electric vector in the $z = 0$ plane rotates in a *clock-wise sense*, and in a conventional way this is called *right-circularly polarized light* (RCP).

Next, we consider the spatial trace of the electric vector along the z-axis *at a certain time*, e.g., at $t = 0$. From Eqs. (4) and (5), it can be seen that the trace of RCP is right-handed as shown in Fig. 2. Moving along the positive z-axis, we find that the

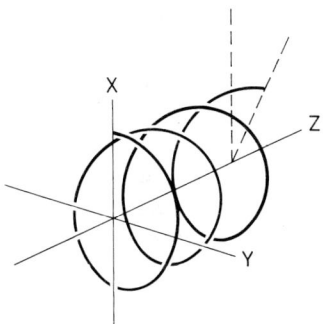

Fig. 2. Instantaneous snapshots ($t = 0$) of the right-circularly polarized light travelling in the right-handed Cartesian frame

electric vector rotates first in the positive y-direction at $z = \pi c/2\omega$ and then in the negative x-direction at $z = \pi c/\omega$. The instantaneous snapshots of the electric vector trace out a *right-handed helix*.

If we choose

$$\vec{u} = (1/\sqrt{2})\begin{pmatrix} 1 \\ i \end{pmatrix}, \tag{6}$$

Eq. (2) is converted to the following Eqs. (7) and (8);

$$E_x(t, z) = (E_0/\sqrt{2}) \cos\omega(t - z/c), \tag{7}$$

$$E_y(t, z) = (E_0/\sqrt{2}) \sin\omega(t - z/c). \tag{8}$$

The Eqs. (7) and (8) correspond to *left-circularly polarized light* (LCP).

In order to describe an electro-magnetic wave (light), we need to know the following characteristics:
i amplitude or intensity E_0,
ii frequency ω,
iii direction of propagation z,
iv orientation of the vibration relative to the known axis, x and y,
v the vibrations of (i) ~ (iv) with time, $\vec{u}$.

To specify (iv), either the electric or the magnetic vector can be chosen. Above, we choose the former, since it plays the dominant role in optical measurements. The orientational characteristics of this vector *in time and space* are termed the *polarization* of the electro-magnetic wave (light). If the loci of the tips of the electric vector vibrating along the z-axis are situated in a plane, the wave is said to be *linearly polarized* (or *plane-polarized*). This linearly polarized light is considered as the result of two, right- and left-, circularly polarized components of opposite handness, which propagate exactly *in phase* with one another. If the two oppositely handed circular polarizations travelling along the optical axis have different velocities, this case is referred to as *circularly birefringent*. This circular birefringence results in *optical rotation*.

2.3 Optical Rotation

This section considers circularly polarized waves with equal amplitudes and frequencies but with opposite handness, which are travelling in the direction of increasing z in a right-handness Cartesian frame. At any particular value of z, the electric vector of circularly polarized light is of constant length and rotates with uniform angular velocity along the direction of propagation, z. If the resulting disturbance caused by the simultaneous presence of the LCP and RCP beams, will take place entirely in the xz-plane, i.e., being linearly polarized, their y-components are always equal and opposite cancelling each other out. However, if the two circularly polarized beams enter into a medium of circular birefringence, their y-components are not cancelled. Supposing their velocities in this medium are v_L and v_R (L and R denoting left- and

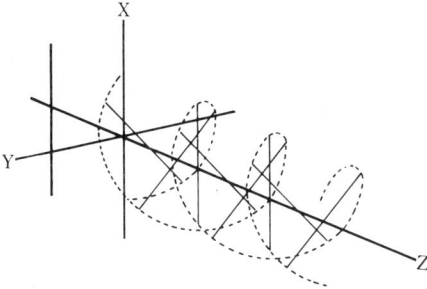

Fig. 3. Direction of vibration of linearly polarized light as a function of z

right-handed) and taking a "snapshot" of the two circularly polarized beams at a certain time, say $t = 0$, the loci of the tips of the electric vector in space appear to be two helices. The pitches of the two helices are different because of the different velocities of propagation in a birefringent medium. Hence with $z = z_0$ the position angles (Φ_L, Φ_R) of the two vectors at this particular time are:

$$\Phi_L = (-2\pi z_0/\lambda)(c/v_L), \qquad \Phi_R = (2\pi z_0/\lambda)(c/v_R). \tag{9}$$

Mathematical definition yields the clockwise handness toward a positive z-direction as positive and vice versa, like angular momentum. As time proceeds, the two vectors at z_0 contrarotate and combine to give a linear polarization halfway between them, i.e., at the angle position:

$$\Phi = (1/2)(\Phi_R - \Phi_L) + \Phi_L \tag{10}$$

$$= (\pi z_0 c/\lambda)[(1/v_R) - (1/v_L)]. \tag{11}$$

Thus, when linearly polarized light travels through a circularly birefringent material, it remains linearly polarized, but its direction of vibration is a function of z (see Fig. 3). The type of helix, left- or right-handed, spatially described by the linear polarization vector, is determined by the sign of the term $[(1/v_R) - (1/v_L)]$. The observer viewing a right-handed helix in the z-direction, sees it constantly rotating clock-wise around the light source ($z = 0$) in the case of Fig. 2. Stated differently, the change of the polarization direction is observed in the order of 7, 6, and so on (see the numerical order in Fig. 4). If the plane of polarization traces out a right-handed helix,

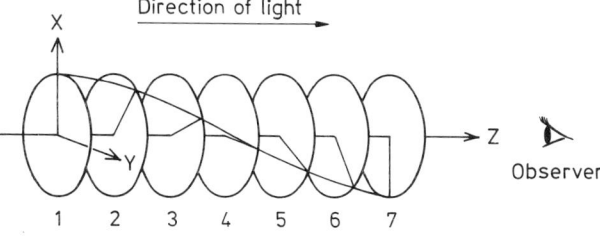

Fig. 4. Change of the polarization direction as viewed by an observer

$v_R > v_L$, as shown in Fig. 2, the medium is said to be *dextrorotatory*. This definition has been used by most physicists [13].

Here it should be emphasized that chemical definition is exactly opposite. According to the convention in physics, the signs coincide with those in mathematics. The traditional chemical signs are converted, however. Thus, the sign in Eqs. (10) and (11) must be converted, as follows:

$$\Phi = (\pi z_0 c/\lambda)[(1/v_L) - (1/v_R)] . \tag{12}$$

The signs of optical activity or of the Faraday effect (magnetically induced optical activity; see Appendix) used by physicists are frequently opposite to the chemically defined ones. Furthermore, the handness in liquid crystals, such as cholestric or chiral smectic ones, often has been defined erroneously and thus confused.

Rewriting the expression for Φ in terms of refractive indeces n_L and n_R, we obtain:

$$\Phi = (\pi z/\lambda)[n_L - n_R] , \tag{13}$$

as $n = c/v$. If $z = 1$ cm, then Φ is described in terms of a radian, and can be obtained from:

$$\Phi = (\pi/\lambda)[n_L - n_R] \quad \text{(radian cm}^{-1}\text{)} . \tag{14}$$

Experimentally, one can use degrees as the rotation unit, and decimeters for the optical path length; then the experimentally obtained rotational angle α is defined as follows:

$$\alpha = \Phi \times 10(180/\pi) = (1800/\lambda)(n_L - n_R) \quad \text{(degree)} . \tag{15}$$

For the rotation in a given solution, we can use the *specific rotation* $[\alpha]$ or the molar rotation $[\Phi]$:

$$[\alpha] = \alpha/dc \tag{16}$$

$$[\Phi] = M[\alpha]/100 , \tag{17}$$

where α is the rotation in degrees, d the path length in decimeters, c the concentration of the given chiral species in g cm^{-3}, and M the molecular weight. The value of Φ is not proportional to the wavelength of the observing light and exhibits a dispersion depending on the wavelength. This phenomenon is known as *optical rotatory dispersion* (ORD). The solvent effect on the refractive index can be corrected by the Lorentz factor $3/(n^2 + 2)$, where n is the refractive index of the used solvent and is variable depending on the wavelength. Therefore, instead of $[\Phi]$, the reduced rotation $[m']$ is used:

$$[m'] = [3/(n^2 + 2)][\Phi] = [3/(n^2 + 2)]M[\alpha]/100 , \tag{18}$$

where M is often replaced by the averaged molecular weight of the repeating units in biopolymers, such as amino acid residues in proteins or nucleotide units in nucleic acids. For composites such as viruses or membrane-bound proteins, as well as for unknown samples, α should be used.

At a farther distance from the absorption region, the rotation of the molecular system is expressed as a sum of the one-term Drude formula:

$$[m'] = \sum_i \frac{a_i \lambda_i^2}{\lambda^2 - \lambda_i^2}. \tag{19}$$

Expanding this expression in a series $(\lambda^2 - \lambda_0^2)^{-1}$, in which λ_0 has to be defined we obtain:

$$[m'] = \sum_i [a_i \lambda_i^2/(\lambda_i^2 - \lambda_0^2) + a_i \lambda_i^2 (\lambda_i^2 - \lambda_0^2)^2/(\lambda_i^2 - \lambda_0^2)^2$$
$$+ a_i \lambda_i^2 (\lambda_i^2 - \lambda_0^2)^2/(\lambda_i^2 - \lambda_0^2)^3 \ldots]. \tag{20}$$

The series converges rapidly if $\lambda_0 \ll \lambda_i$. Then, Eq. (20) may be written in the form:

$$[m'] = a_0 \lambda_0^2/(\lambda^2 - \lambda_0^2) + b_0 \lambda_0^4/(\lambda^2 - \lambda_0^2)^2 + c_0 \lambda_0^6/(\lambda^2 - \lambda_0^2)^3 + \ldots \tag{21}$$

where

$$a_0 \lambda_0^2 = \sum_i a_i \lambda_i^2, \quad b_0 \lambda_0^2 = \sum_i a_i \lambda_i^2 (\lambda_i^2 - \lambda_0^2),$$

$$c_0 \lambda_0^6 = \sum_i a_i \lambda_i^2 (\lambda_i^2 - \lambda_0^2)^2. \tag{22}$$

If $b_0 = c_0$, then Eq. (21) becomes the one-term Drude formula [Eq. (19)]. If $b_0 \neq 0$, but $c_0 = 0$, then:

$$[m'] = a_0 \lambda_0^2/(\lambda^2 - \lambda_0^2) + b_0 \lambda_0^4/(\lambda^2 - \lambda_0^2)^2. \tag{23}$$

Experiments with poly-α-amino acids or proteins have been in perfect agreement with the above when $\lambda_0 = 212$ nm and $b_0 = -630$. Equation (23) is also called Moffitt's equation or Moffitt-Yang's equation.

2.4 Circular Dichroism

k_L and k_R are the absorbances for LCP and for RCP light, respectively. When plane-polarized light is focused on an absorbing sample in which $k_L \neq k_R$, the transmitted light is no longer plane-polarized.

Adding LCP and RCP components of different amplitudes, yields *elliptically polarized light*. The major axis of the ellipsoid is the sum of amplitudes A_R and A_L, and the minor axis is their difference, as shown in Fig. 5. The *ellipticity* θ is defined as the arctangent of the ratio of the minor axis to the major axis:

$$\theta = \tan^{-1} \frac{A_R - A_L}{A_R + A_L} \quad \text{(radian)}. \tag{24}$$

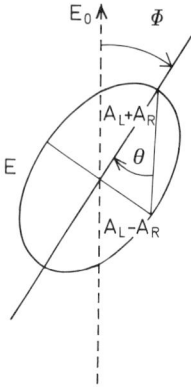

Fig. 5. Elliptically polarized light emerging toward the observer through a circularly dichroic sample. The sign is defined by the chemical convention that Φ is positive for clockwise rotation

Then,

$$\tan \theta = \frac{A_R - A_L}{A_R + A_L} = \frac{\exp(-k_R x/2) - \exp(-k_L x/2)}{\exp(-k_R x/2) + \exp(-k_L x/2)}$$

$$= \frac{1 - \{\exp-(k_R - k_L)x/2\}}{1 + \{\exp-(k_R - k_L)x/2\}} \tag{25}$$

where x is the cell length, and k_R and k_L the absorption coefficients for RCP and LCP, respectively. By the relation of:

$$e^x = 1 + x/1 + x^2/2 + \ldots$$

and very small values of $(k_R - k_L)$, Eq. (25) can be approximated:

$$\tan \theta = \frac{(1/2)(k_R - k_L)x}{2 - (1/2)(k_R - k_L)x} \tag{26}$$

The denominator in Eq. (26) is nearly equal to 2, as $(k_R - k_L)$ is very small. Assuming $(\tan \theta) \cong \theta$, one obtains:

$$\theta \cong (1/4)(k_R - k_L) \quad \text{(radian cm}^{-1}\text{)}. \tag{27}$$

This relation is derived from the physical definition as well as from the optical rotation. By chemical definition, if the plane of polarization as seen by an observer looking toward the light source, is rotated clock-wise, the rotation is positive. Equation (27) then converts to:

$$\theta \cong (1/4)(k_L - k_R). \tag{28}$$

Experimentally, one can use degrees as the rotation unit and decimeters for the optical path length. Consequently, the experimentally obtained ellipticity [θ'], which is called *specific ellipticity*, is described as follows:

$$[\theta'] = \theta \times (180/\pi) \times 10 \times (1/c') \quad \text{(degree dm cm}^3 \text{ g}^{-1}\text{)}, \tag{29}$$

where the concentration c' is expressed in g cm^{-3}. The concentration is usually in moles per liter, in the case of which the extinction coefficients defined per centimeter of path length can be used instead of the absorption coefficients. Thus, Eq. (29) leads to:

$$[\theta] = (18 M/4\pi) \times 2.303 \times (\varepsilon_L - \varepsilon_R) \times (10^3/M)$$
$$= 3300(\varepsilon_L - \varepsilon_R) \quad \text{(degree cm}^{-1}\text{)} \tag{30}$$

where M is the molecular weight of the sample, and ε_L and ε_R are the extinction coefficients for the LCP and RCP light of the sample. In Eq. (30), [θ] is called the *molar ellipticity*; occasionally the reduced ellipticity multiplied by $[3/(n^2 + 2)]$ of the Lorentz correction is also used.

ORD and CD are connected through the Kronig-Kramers integral transform. Thus, for the k-th transition it follows that:

$$[\Phi_k(\lambda)] = (2/\pi) \int_0^\infty [\theta_k(\lambda_i)] \frac{\lambda_i^2}{\lambda^2 - \lambda_i^2} \, d\lambda \tag{31}$$

$$[\theta_k(\lambda)] = -(2/\pi\lambda) \int_0^\infty [\Phi_k(\lambda_i)] \frac{\lambda_i^2}{\lambda^2 - \lambda_i^2} \, d\lambda \, . \tag{32}$$

The Kronig-Kramers relation is of fundamental importance for optics and for physics in general [13]. Here, these equations do not seem practical because of the integration of the wavelength from 0 to ∞. However, these are very useful for calculating the molar ellipticity magnitude from the observed ORD curve [14].

3 Theory of Optical Activity

In 1892, Biot confirmed that the colors on propagating white light parallel to the optical axis of a quartz crystal placed between crossed polarizers arise from two distinct effects, the rotation of the plane of polarization of monochromatic light and dispersion of the rotation with respect to wavelength. Biot's discovery was extended to the optical rotation of natural products in solution or in the liquid phase, and this is of chemical significance, as it indicates that rotation is a molecular effect.

Later, Pasteur [15] had arrived at the general stereochemical criterion for a chiral or dissymmetric molecular structure. Thus, the specific rotations of the two sets of sodium ammonium tartrate crystals in solution, isolated from the racemic mixture by hand-picking, were equal in magnitude and opposite in sign, from which Pasteur inferred that enantiomorphism of the dextro- and laevorotatory crystals is reproduced in the microscopic stereochemistry of the (+)- and (—)-tartaric acid molecules. The term dissymmetry or chirality is used when there is no superimposability between the two enantiomers, as seen in Sect. 2.1.

On the other hand, the stereochemical concept in inorganic chemistry was established by Werner [16], who succeeded in resolving Co(III) complexes, freeing the subject of optical activity from the framework of the asymmetric carbon atom.

The first attempt to formulate a theory of optical rotation in terms of the general equations of wave motion was made by MacCullagh [17]. His theory was extensively developed on the basis of Maxwell's electromagnetic theory. Kuhn [18] showed that the molecular parameters of optical rotation were elucidated in terms of molecular polarizability β connecting the electric moment $\vec{\mu}$ of the molecule, the time-derivative of the magnetic radiation field H, and the magnetic moment m with the time-derivative of the electric radiation field E as follows:

$$\vec{\mu} = \alpha \vec{E} - \frac{\beta}{c} \frac{\partial \vec{H}}{\partial t} \tag{33}$$

$$\vec{m} = \frac{\beta}{c} \frac{\partial \vec{E}}{\partial t} \tag{34}$$

The quantum-mechanical treatment of optical activity was initiated by Rosenfeld [19] who showed that rotatory polarizability β of Eqs. (33) and (34) is represented by:

$$\beta_a = \frac{c}{3\pi h} \sum \frac{\text{Im}\{\langle a|\vec{\mu}|b\rangle \cdot \langle b|\vec{m}|a\rangle\}}{v_{ab}^2 - v^2}, \tag{35}$$

where the expression has been simplified to take account of the fact that the molecules are *oriented at random*. In Eq. (35), v is the light frequency and v_{ab} the frequency associated with a transition from state a to state b. The symbol $Im\{\ \}$ means '*imaginary part of*' in the sence:

$$Im\{u + iv\} = v$$

if u and v are real. Subscript a on β means that the formula gives the appropriate value of β for molecules in state a. The matrix components of the electric moment of the molecule are denoted by $\langle a|\vec{\mu}|b\rangle$ and those of the magnetic moment by $\langle b|\vec{m}|a\rangle$, where

$$\vec{\mu} = \sum_i e\vec{r}_i, \quad \vec{m} = \sum_i (e/2mc)(\vec{r}_i \times \vec{p}_i + 2S_i), \tag{36}$$

the sum being extended over all the electrons in the molecule, $\vec{r}_i$ being the position vector, $\vec{p}_i$ the momentum vector, and $\vec{S}_i$ the spin angular momentum vector of the i-th electron. Then, the theory of optical activity has been reverted to the calculations of the electronic parameters such as $\vec{\mu}$ and $\vec{m}$ by using appropriate models for the calculation.

3.1 Symmetry Considerations

The theory of optical activity would be understood in terms of symmetry considerations at the first stage. The elements of symmetry are the geometric elements in relation to which the symmetry operations are carried out, and are classified in the following:

i   plane of symmetry σ,
ii axis of symmetry C_n,
iii improper axis of symmetry S_n,
iv center of symmetry i.

The plane of symmetry σ can divide into two parts in such a way as one is the image of the other. For example, cyclopropane possesses three vertical planes of symmetry σ_v and one horizontal plane of symmetry σ_h. The *axis of symmetry* C_n can define a rotation of the molecule around the axis, if the rotation through $2\pi/n$ gives an equivalent configuration. Thus, Fig. 6 shows that a rotation of $2\pi/3$ around the vertical axis (C_3 axis) of the cyclopropane ring permutes each corner by $2\pi/3$ to its neighbor. A rotation around the axis (C_2 axis) intercepting the cyclopropane ring to two identical parts can interchange the two hydrogen atoms at the corner carbon atom. An *improper axis of symmetry* S_n can result from a rotation of order n followed by a reflection in a plane perpendicular to the rotation axis. Now, let us consider ethane in a staggered configuration (Fig. 7.). Although the symmetry in the molecule does not include σ and C_6, the combination of C_6 and σ gives an identical configuration. This operation is called S_6.

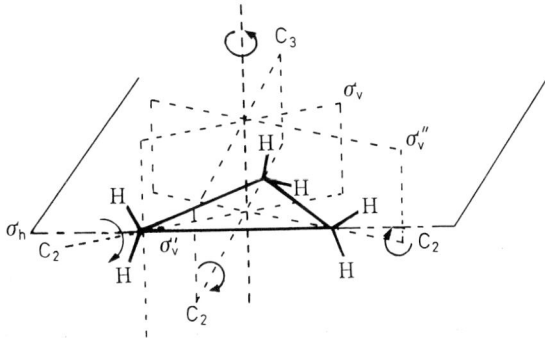

Fig. 6. Rotation of $2\pi/3$ around the vertical axis (C_3 axis) to the cyclopropane ring

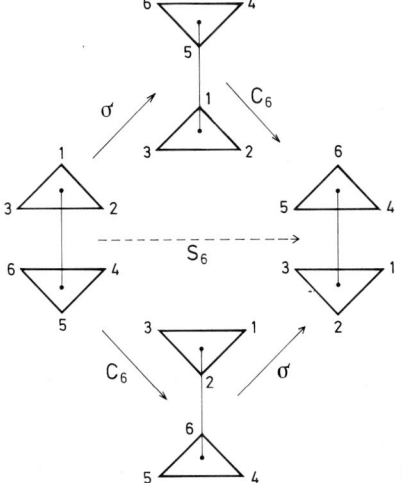

Fig. 7. Staggered configurations in ethane

If a molecule possesses a proper axis, C_n, as well as a twofold axis (C_2) perpendicular to the proper axis, the point group is denoted as D_n. By adding both the horizontal plane and the set of n vertical planes to C_n, a group called D_{nh} is generated. The combination of D_n with a set of dihedral planes bisecting between adjacent pairs of C_2 leads to the symmetry D_{nd}. If the symmetry elements C_n and vertical planes (σ_v) intersecting the axis are present, the molecule belongs to the point group C_{nv}. If the molecule has C_n symmetry and a symmetry plane vertical to C_n, it belongs to C_{nh}.

A molecule A is chiral, when its mirror image B is not superimposable on A. Only the groups C_1, C_n, D_n are compatible with *chirality*. Especially the molecules belonging to the C_1 group are called *asymmetric*. The molecules having C_n or D_n symmetry are also *chiral* and have an *axial symmetry*.

When the symmetry of a given molecule has been determined, electronic transitions fall into the following three categories:

i a. electric allowed — magnetic forbidden
    b. magnetic allowed — electric forbidden
    [Groups C_i, C_{nh}, D_{nh} ($n \neq 2$), S_{2n} (n, odd), O_h, T_d]

ii electric and magnetic allowed, but perpendicular
    [Groups C_s, C_{nv}, D_{2d}, S_{2n} (n, even)]

iii Electric and magnetic allowed, parallel
    (Groups C_n, D_n).

For the third group one can expect inherent optical activity. However, examples of this case are rare. Hexahelicene (5 in Table 1) is the best example. The almost all-optical activity theory has been concerned with the first group, in which a perturbation theory to the zero-th order function describing either an electric or magnetic-allowed transition has been used. *Induced optical activity* or *induced circular dichroism* (ICD) is generated through the mechanism of the first group. Table 1 shows many examples of achiral and chiral molecules together with their point groups.

3.2 Coupled Oscillator Mechanism

For the optical activity of achiral chromophores with a dissymmetric environment, two types of theoretical treatments have been proposed: coupled oscillator treatment and one-electron treatment. The charge distribution of the magnetic dipole transition correlates Coulombically with an electric dipole induced in the substituents, and the colinear component of the induced dipole provides, with the zero-th order magnetic moment, a non-vanishing rotational strength.

The coupled oscillator mechanism involves coupling between the transition moments of two adjacent chromophores; these must have a spatial relationship in which the interacting moments are non-parallel. (If the moments are parallel, only the absorption spectrum is affected; hyperchromism is shown if the chromophores are arranged in a head-to-tail mode, and hypochromism if they are stacked one over the other in the head-to-head mode [20]. By knowing the optical activity, it is possible to deduce the relative configuration between the given chromophores, and vice versa.

Let us consider the optical activity of calycanthine [21,22] as an example (Fig. 8). The absolute configuration in calycanthine was determined by X-ray crystallographical analysis [23]. However, it should be emphasized that the X-ray data need to be

Table 1. Several examples of point groups and symmetry elements

Symbol of group	Elements of Symmetry	Examples
C_1		1, 2
C_s	σ	3, 4
C^z		5 (Hexahelicene), 6
$S_2 \equiv C_i$	i	7
C_{2v}	C_2, σ_v	8
C_{2h}	C_2, σ_h	9

Table 1 (Continued)

Symbol of group	Elements of Symmetry	Examples
C_2	C_2	10
D_{2h}	C_2, σ_h	11
D_{3d}	C_3, C_2, σ_v	12
D_{6h}	C_6, C_2, σ_h	13
T_d	C_3, C_2, σ	14
S_4	C_2	15
O_h	C_4, C_2, S_4	16
C_2		17 (Tröger base)

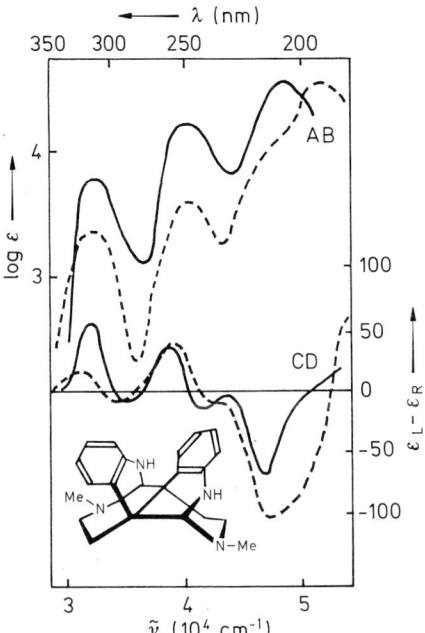

Fig. 8. Absorption spectra (AB; upper curves) and circular dichroism (CD; lower curves) of calycanthine (formula in the inset of the figure). The solid lines refer to the experimental spectra and the broken lines to the theoretical curves for the configuration illustrated, calculated by the Pariser-Parr-Pople method using the dipole velocity procedure[zl, zz)]

verified. Harada et al. [24)] reported on the absolute stereochemistry of 3-epicaryoptin, caryoptin, and clerodin, and found that the absolute configurations of these compounds by the optical activity disagree with those derived from their X-ray data. More recently, the X-ray data for the compounds were revised by Rogers [25)], whose analyses led to the same conclusion as that of Harada.

Reconsidering the optical activity of calycanthine, we shall focus on two chromophores which maintain the pertinent distance and geometry. It is assumed that the electronic states of these chromophores do not mix with those of other chromophores. The linear combinations of the locally excited configurations were taken to be:

$$\Psi_{\pm} = (1/\sqrt{2})(\Psi_{a0} \pm \Psi_{0b}) \tag{37}$$

where

$$\Psi_{00} = \psi_0^{(1)}\psi_0^{(2)} \tag{38}$$

$$\Psi_{0b} = \psi_0^{(1)}\psi_b^{(2)} \tag{39}$$

$$\Psi_{a0} = \psi_a^{(1)}\psi_0^{(2)} \tag{40}$$

Suffix 0, a, and b are the ground state, and the singly excited states of chromophores (1) and (2). In Eq. (37), the plus and minus signed wave function Ψ_+ and Ψ_- correspond to A and B symmetries, respectively, assumed to be a C_{2v} point symmetry for the group of the chromophores (1) and (2).

In the A-symmetry transition $\Psi_+ \leftarrow \Psi_{00}$, the electric and magnetic transition moments can be described as:

$$\vec{\mu}(\Psi_+ \leftarrow \Psi_{00}) = \langle \Psi_{00} | e\vec{r} | \Psi_+ \rangle$$
$$= (1/\sqrt{2}) \langle \psi_0^{(1)} \psi_0^{(2)} | e\vec{r} | (\psi_a^{(1)} \psi_0^{(2)} + \psi_0^{(1)} \psi_b^{(2)}) \rangle \quad (41)$$

$$\vec{m}(\Psi_+ \leftarrow \Psi_{00}) = \langle \Psi_+ | \vec{r} \times \vec{p} | \Psi_{00} \rangle$$
$$= (e/2mc) \{ \langle \Psi_+ | \vec{r}_0 \times \vec{p} | \Psi_{00} \rangle + \langle \Psi_+ | \vec{a} \times \vec{p} | \Psi_{00} \rangle \} \quad (42)$$

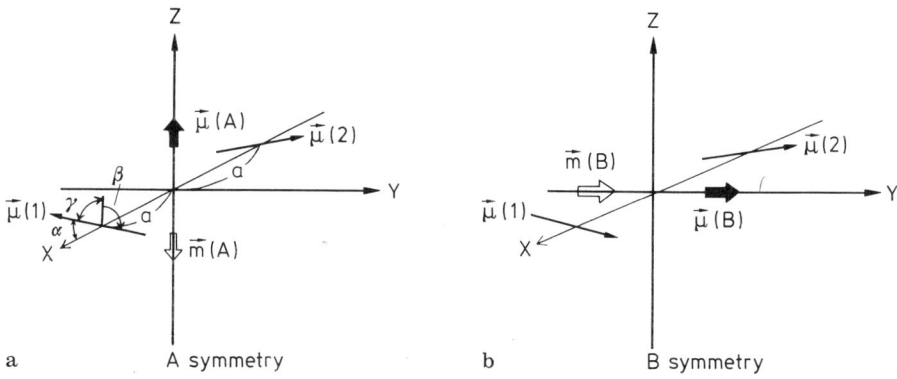

Fig. 9. The combination of two transition dipoles in the framework of C_{2v} point group

If the origin of the group is situated at the midpoint between the centers of two chromophores, the vector $\vec{r}$ is described as the sum of the vector of the center $\vec{a}$ and the vector of the electron in the chromophore exhibited by the local coordinate $\vec{r}_0$ (Fig. 9). From the orthogonality of the wave-functions of each chromophore and unvariable condition of $\vec{a}$, Eqs. (41) and (42) are reduced to:

$$\vec{\mu}(\Psi_+ \leftarrow \Psi_{00}) = (1/\sqrt{2}) \{ \langle \psi_0^{(1)} | e\vec{r}_0 | \psi_a^{(1)} \rangle + \langle \psi_0^{(2)} | e\vec{r}_0 | \psi_b^{(2)} \rangle \}$$
$$= (1/\sqrt{2}) \{ \vec{\mu}(1) + \vec{\mu}(2) \} \quad (43)^1$$

$$\vec{m}(\Psi_+ \leftarrow \Psi_{00}) = (i\pi) \{ \tilde{v}_{0a} [\vec{a} \times \vec{\mu}(1)] + \tilde{v}_{0b} [\vec{a} \times \vec{\mu}(2)] \} \quad (44)$$

$$\vec{m} = (e/2mc) [\langle \psi_a^{(1)} | \vec{a} \times \vec{p} | \psi_0^{(1)} \rangle + \langle \psi_b^{(2)} | \vec{a} \times \vec{p} | \psi_0^{(2)} \rangle].$$

The momentum operator $\vec{p}$ can be converted to:

$$\langle \psi_a | p_x | \psi_0 \rangle = i \, 2\pi m v_{0a} \langle \psi_a | x | \psi_0 \rangle,$$

---

1 Note: In deriving Eq. (43), the following relations are used. The first term of the right side in Eq. (42) is corresponding to the transition magnetic moment of the single chromophore, and then is neglected for the common chromophores exhibiting absorption in the ultraviolet or visible region, in comparison with those of the chromophore couple.

resulting in:

$$(e/2mc) \langle \psi_a | \vec{a} \times \vec{p} | \psi_0 \rangle = (i\pi\tilde{v}_{0a}) \langle \psi_a | \vec{a} \times \vec{p} | \psi_0 \rangle$$
$$= i\pi\tilde{v}_{0a} \langle \psi_a | \vec{a} \times e\vec{r}_0 | \psi_0 \rangle,$$

where $\tilde{v} = v/c$.

For identical chromophores, $\tilde{v}_{0a} = \tilde{v}_{0b}$. This leads to the following equation:

$$\vec{m}(\Psi_+ \leftarrow \Psi_{00}) = i\pi\tilde{v}_{0a}\{\vec{a} \times \vec{\mu}(1) + \vec{a} \times \vec{\mu}(2)\}. \tag{45}$$

As shown in Eq. (35), the quantum mechanical foundation expresses the following equation for rotational strength of $0 \to a$ transition:

$$R_{0a} = (\pi\tilde{v}_{0a}/2) \operatorname{Im} \{\vec{\mu}_{0a} \cdot \vec{m}_{a0}\}. \tag{46}$$

This equation has been developed by Rosenfeld [19].

From Eqs. (41) and (45), one can obtain:

$$R(\Psi_+ \leftarrow \Psi_{00}) = (\pi\tilde{v}_{0a})\{(1/\sqrt{2})(\vec{\mu}(1) + \vec{\mu}(2))\} \cdot \{\vec{a} \times \vec{\mu}(1) + \vec{a} \times \vec{\mu}(2)\}. \tag{47}$$

The excitation dipole, $\vec{\mu}$, has three components which are directed respectively parallel to the z and y axis, and along the x axis which is defined by the line of length 2a joining the centers of the dipoles, as shown in Fig. 9. The magnitudes of the components are determined by the cosines of the vertical, tangential, and radial angles, γ, β, and α, respectively, between the excitation dipole and the local Cartesian axes (see Fig. 9.).

If $|\vec{\mu}(1)| = |\vec{\mu}(2)|$, and the vectors are described by the above-mentioned framework of the Cartesian, Eq. (47) can be simplified. The coupling modes of the excitation dipoles with B and A symmetry give rise to a right- and left-handed charge displacement, respectively, for the depicted absolute configuration (Fig. 9). Assuming the unit vectors along the x-, y-, and z-axes in the Cartesian to be $\vec{i}, \vec{j}$, and $\vec{k}$, one can describe $\vec{\mu}(1)$ and $\vec{\mu}(2)$ in Eq. (47) as:

$$\vec{\mu}(1) = \mu(\vec{i} \cos \alpha + \vec{j} \cos \beta + \vec{k} \cos \gamma) \tag{48}$$

$$\vec{\mu}(2) = \mu(-\vec{i} \cos \alpha - \vec{j} \cos \beta + \vec{k} \cos \gamma). \tag{49}$$

Then,

$$\{\vec{\mu}(1) + \vec{\mu}(2)\} = 2\mu\vec{k} \cos \gamma \tag{50}$$

and

$$\{\vec{a} \times \vec{\mu}(1) + \vec{a} \times \vec{\mu}(2)\} = a\{\vec{i} \times \vec{\mu}(1) - \vec{i} \times \vec{\mu}(2)\} \tag{51}$$
$$= a\mu\{\vec{k} \cos \beta + \vec{j} \cos \gamma)$$
$$- (-\vec{k} \cos \beta + \vec{j} \cos \gamma)\} \tag{52}$$
$$= 2a\mu\vec{k} \cos \beta. \tag{53}$$

Equations (50) and (53) are inserted into Eq. (47) leading to:

$$R(\Psi_+ \leftarrow \Psi_{00}) = (\pi\tilde{\nu}_{0a})\{(1/\sqrt{2})\,2\mu\vec{k}\cos\gamma\} \cdot \{d\mu\vec{k}\cos\beta\}$$
$$= \sqrt{2}\pi\tilde{\nu}_{0a}\mu^2 d\cos\beta\cos\gamma\,, \tag{54}$$

where $2a = d$.

Furthermore, $R(\Psi_- \leftarrow \Psi_{00})$ is obtained similarly to the case of $R(\Psi_+ \leftarrow \Psi_{00})$, as follows:

$$R(\Psi_- \leftarrow \Psi_{00}) = -\sqrt{2}\pi\tilde{\nu}_{0a}\mu^2 d\cos\beta\cos\gamma\,. \tag{55}$$

Equations (54) and (55) correspond to the rotational strengths of the A and B symmetry, respectively. Thus, the rotational strengths of the two resultant transitions are equal in magnitude and opposite in sign (see Table 2).

When two chromophores with the dipoles $\vec{\mu}(1)$ and $\vec{\mu}(2)$, separated by the vector $\vec{r}_{12}$, are interacting, the interaction energy V_{12} is expressed by:

$$V_{12} = (1/d^3)\{\vec{\mu}(1)\cdot\vec{\mu}(2) - 3[\vec{\mu}(1)\cdot\vec{r}_{12}]\,[\vec{\mu}(2)\cdot\vec{r}_{12}]/d^2\}\,. \tag{56}$$

The separation of the energy, $\Delta\tilde{\nu}$, for two transitions with A and B symmetry in the interacting chromophores is:

$$\Delta\tilde{\nu} = \tilde{\nu}(A\text{ symmetry}) - \tilde{\nu}(B\text{ symmetry})$$
$$= \{V_{12}(A\text{ symmetry}) - V_{12}(B\text{ symmetry})\}/h\,. \tag{57}$$

V_{12} (A symmetry) and V_{12} (B symmetry) can be expressed by:

$$V_{12}(A\text{ symmetry}) =$$
$$(1/d^3)\{\mu^2(\vec{i}\cos\alpha + \vec{j}\cos\beta + \vec{k}\cos\gamma)\cdot(-\vec{i}\cos\alpha - \vec{j}\cos\beta + \vec{k}\cos\gamma)$$
$$-(3/d^2)\mu^2(\vec{i}\cos\alpha + \vec{j}\cos\beta + \vec{k}\cos\gamma)\cdot(\vec{i}d)(-\vec{i}\cos\alpha - \vec{j}\cos\beta$$
$$+ \vec{k}\cos\gamma)\cdot(\vec{i}d)\} = (\mu^2/d^3)(2\cos^2\alpha - \cos^2\beta + \cos^2\gamma) \tag{58}$$

$$V_{12}(B\text{ symmetry}) = -(\mu^2/d^3)(2\cos^2\alpha - \cos^2\beta + \cos^2\gamma)\,. \tag{59}$$

Then,

$$\Delta\tilde{\nu} = (2\mu^2/hd^3)(2\cos^2\alpha - \cos^2\beta + \cos^2\gamma)\,, \tag{60}$$

where d is equal to $2a$ in Fig. 9.

If it is assumed that the excitation moments of the two aniline units in calycanthine are point dipoles located at the centers of the benzene rings, the distance d is 4.14 Å and the vertical, tangential, and radial cosine are $+0.760$, -0.469, and $+0.448$, respectively, for the α-band (310 nm band), and $+0.404$, $+0.883$, and $+0.238$, respectively, for the p-band (250 nm). These values result from the absolute configuration determined by X-ray, strengths and frequency separations for the α- and p-bands,

Table 2. Circular dichroism and absorption spectra of calycanthine and the theoretical rotational strengths and frequency separations [21]

Band	Component	R × 10⁴⁰ (c.g.s.) (calc.)	$\Delta\tilde{v}$ (cm⁻¹)	$\tilde{v}_{CD}$ (cm⁻¹)	R × 10⁴⁰ (c.g.s.)	ε_{max} (M⁻¹cm⁻¹) $\tilde{v}_{max}$ (cm⁻¹)	
α	B	+ 46	+350	31 500	+38	6 300	32 300
	A	− 46		34 500	−7		
p	A	+178	−800	38 500	+24	18 600	39 600
	B	−178		41 600	−30		

as tabulated in Table 2. Some deviations of the theoretical from the experimental values are due to the extent of mixing of transitions with the same symmetry. This model has been used by many authors to determine the absolute configurations of various aromatic compounds. Recently, Nakanishi and Harada [26] emphasized that the analysis of the frequency and sign order of the two oppositely-signed CD bands, based on the isoenergetic transitions, is quite useful for the analysis of the absolute configuration of natural products such as steroid diols. Thus, chiral diols from benzoate diesters give a characteristic bisignate CD couplet in the 220–280 nm region. The sign of the CD band at the longer wavelength side indicates the chirality around the dibenzoate moiety (Fig. 10).

However, it is necessary to determine precisely the orientation of the dipole moments, and to ensure that the observed CD is not the resultant from a mixture of conformers. Moreover, even with a correct choice of the dipoles, there are often shortcomings. For instance, studies of 1,1'-spirobisindanes have shown substantial discrepancies between the theoretical and experimental results for several products of known configuration [27]. It has been shown that the signs of the bands stand in direct relation to the configuration of the spiro-atoms, which means that the optical activity is really governed by the coupling mechanism including other types of interaction.

Triptycene derivatives are more interesting in clarifying the relation between their configurations and optical activity. The results published by Tanaka [28] were not in

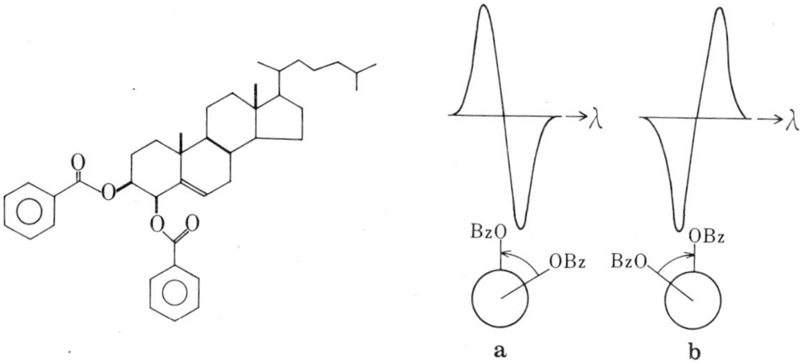

Fig. 10a and b. Schematic CD spectral features and chirality of α-glycol dibenzoates. **a** Negative chirality; **b** Positive chirality

agreement with X-ray determinations by means of the Bijvoet method. This led Tanaka to reconsider the Bijvoet theory and to propose a revision of the Fischer convention. However, recent experiments on the opposite (polar) faces of non-centro-symmetric crystals, such as ZnS and CdS, by noble gas ion-reflection mass spectrometry seem to invalidate Tanaka's proposal [29]. Moreover, Mason predicted the importance of origin dependency on the calculation of rotational strength for (—)-1,5-diamino-9,10-dihydro-9,10-ethanoanthracene which has a similar structure to that of triptycene, and he showed that using the dipole-velocity procedure, instead of the dipole-length procedure, affords a concordant configuration assignment [30]. Kaito, Tajiri, and Hatano selected the origin at the center of the charge-distribution of the ground state for calculating the optical activity of various kinds of triptycene derivatives using a type of molecules-in-molecule method calculation [31]. Our calculations based on X-ray crystallographical analyses using the Bijvoet method are consistent with the rotational strengths in their magnitudes and in signs obtained experimentally. In calculations of rotational strength by use of the Bijvoet model, one should ensure no or less charge-transfer contribution to the optical activity as well as a mixing of a given state with other electronic states.

Next, consider the simplest model in which $\vec{\mu}(1)$ and $\vec{\mu}(2)$ are perpendicular to the line joining them. The only geometric variables are the distance between the dipoles, d, and the dihedral angle between the dipoles, α. The rotational strength R and dipole strength D, for a given transition of $0 \rightarrow a$ with the wave-number $\tilde{v}_{0a}$ are given as:

$$R_{\pm} = \pm (\pi \tilde{v}_{0a}/2) \, d\mu^2 \sin \alpha \tag{61}$$

$$D_{\pm} = \mu^2 (\pm \cos \alpha) \tag{62}$$

The signs of plus and minus are corresponding to the in-phase (B symmetry) and out-of-phase (A symmetry) mode couplings of interacting electric dipole transitions (see Fig. 9). The dihedral angle is signed plus when the $\vec{\mu}(2)$ vector rotates in a clockwise sense as viewed from the $\vec{\mu}(1)$ vector to the $\vec{\mu}(2)$. There is a considerable cancellation of rotational strengths because of the heavy overlapping of the peaks. In general, the splitting energy calculated with the dipole-dipole interaction approximation is about five times larger than the observed value.

A more accurate method for the estimation of the dihedral angle α is expected by using:

$$\tan^2 (\alpha/2) = (\mu_-/\mu_+). \tag{63}$$

From this method we estimated the dihedral angles of anions and dianions of chiral bianthryl derivatives [32].

The experimental dipole strength and rotational strength are obtained in c.g.s. units from the appropriate band areas by means of the following expressions:

$$D = 9.184 \times 10^{-39} \int \varepsilon(\tilde{v}) \, \tilde{v} \, d\tilde{v} \qquad \text{c.g.s. unit} \tag{64}$$

$$R = 2.296 \times 10^{-39} \int [(\varepsilon_L - \varepsilon_R) \, \tilde{v}] \, d\tilde{v} \qquad \text{c.g.s. unit}, \tag{65}$$

where wavenumber $\tilde{v}$ is in cm^{-1} and the molar extinction coefficient ε is in M^{-1} cm^{-1} or dm^3 mol^{-1} cm^{-1}.

3.3 The Asymmetrically Perturbed Field Mechanism

The one-electron mechanism arises from the electric and magnetic moments of a transition being perturbed by a static field generated by the asymmetric surrounding group(s) in the given molecule. Hence the mechanism is called "*asymmetrically perturbed field mechanism*". The $\vec{\mu} \cdot \vec{m}$ mechanism enables the electrically-allowed transition of one group to interact with other magnetically-allowed transitions of a neighboring group. These mechanisms clearly depend on the geometry of the given chromophore and asymmetric surroundings.

Now lets consider the rotational strength of the n-π^* transition for one example generating the optical activity through the mechanism. Thus the rotational strength is given by:

$$R_{n\pi}^* = \text{Im}\{\langle n| \vec{\mu} |\pi^*\rangle \cdot \langle \pi^*| \vec{m} |n\rangle\} \tag{66}$$

For optically active (chiral) carbonyl compounds, the electrostatic distortion around the carbonyl group has been considered on the basis of the *one-electron approximation*[33]. This electrostatic distortion is due to an electrostatic field which is caused by incomplete shielding of the nuclei of the molecule. The potential perturbation has been expressed as the mixing $2p_{yz}$ with $3d_{yz}$ orbitals[34]. These perturbation approaches have been demonstrated for many saturated ketones, and provide good agreement between experimental and calculated data[35]. In the molecular orbital theory, the chromophore is no longer considered as a symmetric entity, when the chromophore is perturbed by a chiral environment. Many molecular orbital approximations have been used in calculations: the extended Hückel method[36], the SCF-MO method[37], and the CNDO/S MO method[38]. Although these theoretical approximations have been examined successfully, the analyzed examples are limited to rather simple cases. Recently, an *ab initio* (from the beginning) localized orbital analysis of a saturated ketone, using the random-phase approximation, has been examined to clarify the optical activity[39]. The calculated energy and rotational strengths for the n-π^* transition are in excellent agreement with experimental results. This paper[39] includes the detailed formalism and parameters together with the problems concerning the higher-lying bands and origin dependency. The *ab initio* calculation predicts occasionally higher excitation energies. It is necessary to check the validity of the wave functions in comparison with experimentally obtained ones.

Here we emphasize the importance of the semi-empirical geometric rules on the rotational strength, especially its sign, arising from the one-electron mechanism. Moffitt[40], using a coordinate system coinciding with symmetry planes and a nodal surface of the n-π^* transition, proposed that the sign of the contribution to the optical activity varies as the simple product x, y, and z of its coordinates, as illustrated in Fig. 11. The region around the carbonyl group was divided into eight sectors with the sign distribution given in Fig. 11. The *octant rule* is strongly confirmed by the *one-electron theory*[41]. Various factors for consideration in assembling data with the octant rule have been summarized by Klyne and Kirk for saturated ketones[42] and by Snatzke and Snatzke-Zamojska for unsaturated ketones and lactones[43]. The octant rule allows the sign of the contribution of substituents to be determined. Although the octant rule has been established and used for the chiral recognition of many compounds, much caution must be taken before it is used in interpreting experimental data. Dividing the chiral molecule into different spheres is often a better approach to use the octant rule, because it helps to avoid incorrect assignments [44]. The validity of the rule is reduced when the balance between positive and negative contributions results in a very small Cotton-effect or when the major part of the molecule lies in one of the symmetry planes. The effects of substituents with lone paired electrons can only hardly explain the signs by means of a single rule when the relative orientation between the lone paired electrons and the given chromophore plays an important role. In conclusion, the octant rule must be applied to a series of analogous compounds, and it is misleading to use it when no model or no reference is available on which its applicablity is verified. Kagan [45] summarized the applicability and limitation of the octant rule for various kinds of carbonyl, nitroso, azo, and azoxy compounds.

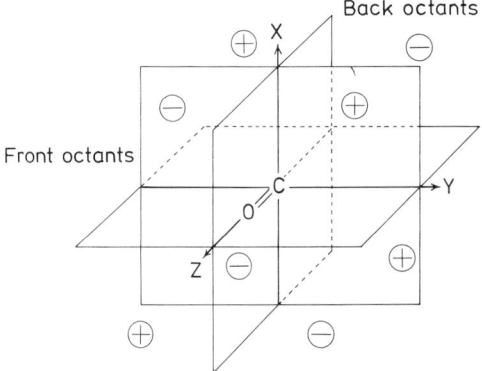

Fig. 11. Octant distribution for the carbonyl group of a ketone

3.4 Induced Optical Activity

Optical activity arises from the coupling of given electric-allowed transitions with a chiral orientation (coupled oscillator mechanism or two-electron mechanism) or from the electric or magnetic moments of a transition being pertubed by a chiral static field (asymmetrically perturbed field mechanism or one-electron mechanism) in the given *one molecule*. A similar mechanism of the optical activity can be expected for *molecular assemblies* which are composed of chiral and achiral ones. This type of optical activity is called "*induced optical activity*" and depends on types of intermolecular interaction modes.

The observed induced circular dichroism (ICD) is classified as follows:
i   dispersion force-induced circular dichroism (DICD),
ii   hydrogen bonding-induced circular dichroism (HBICD),
iii   ionic coupling-induced circular dichroism (ICICD),
iv   ligation-induced circular dichrosim (LICD),
v   charge-transfer-induced circular dichroism (CTICD),
vi   liquid crystal-induced circular dichroism (LCICD),
vii   hydrophobic interaction-induced circular dichroism (HIICD).

It is well known that optical activity is induced for achiral species in the presence of chiral species, giving rise to ICD for the absorption band(s) of the achiral species. In the introduction Section, a historical summary is described in brief. The theory of ICD has been developed by Tinoco, Jr. [46], Mason [47], and Schipper [48].

Now, let us consider a system where an achiral molecule (A) and a chiral molecule (C) have a fixed mutual orientation. An electronic transition of the achiral molecule from the ground state $|A_0\rangle$ to the excited state $|A_a\rangle$, higher in energy by E_{0a}, has a zero-order (non-perturbed) electric dipole moment $\vec{\mu}_{0a}$ and an orthogonal magnetic dipole moment $\vec{m}_{a0}$. These moments are increased in the molecular pair (A–C) by first-order dynamic coupling as:

$$\vec{\mu}_{0a}^{(1)} = -\sum_c 2\vec{\mu}_{0c}\langle A_0 A_a | V | C_0 C_c \rangle E_{0c}/(E_{0c}^2 - E_{0a}^2) \tag{67}$$

$$\vec{m}_{a0}^{(1)} = -\sum_c 2[\vec{m}_{c0} + \pi i \tilde{\nu}_{0c}(\vec{R}_{AC} \times \vec{\mu}_{0c})]$$
$$\times \langle A_0 A_a | V | C_0 C_c \rangle E_{0a}/(E_{0c}^2 - E_{0a}^2), \tag{68}$$

where $\vec{\mu}_{0c}$ and $\vec{m}_{c0}$ refer to the electric and the magnetic moments of the transition with a wavenumber $\tilde{\nu}_{0c}$ of the chiral molecule. Here, one considers the transition from the ground state $|C_0\rangle$ to an excited state $|C_c\rangle$ of the chiral molecule. The potential operator V refers to the intermolecular Coulombic potential between A and C, and becomes:

$$\langle A_0 A_a | V | C_0 C_c \rangle = \vec{\mu}_{0a} \cdot \vec{\mu}_{0c} G_{AC}/R_{AC}^3, \tag{69}$$

where R_{AC} is the distance between the centers of A and C, and G_{AC} is the angular factor for the dipole-dipole potential between $\vec{\mu}_{0a}$ and $\vec{\mu}_{0c}$.

The zero- and first-order moments give the first-order rotational strength $R_{0a}^{(1)}$ and the second-order rotational strength $R_{0a}^{(2)}$, which are induced for the given transition $|A_a\rangle \leftarrow |A_0\rangle$ of the achiral molecule:

$$R_{0a}^{(1)} = \sum_c \frac{2\langle A_0 A_a | V | C_0 C_c \rangle}{(E_{0c}^2 - E_{0a}^2)}$$
$$\times \{E_{0a}[\pi \tilde{\nu}_{0c} \vec{R}_{AC}(\vec{\mu}_{0a} \times \vec{\mu}_{0c}) + i\vec{\mu}_{0a} \cdot \vec{m}_{c0}] + iE_{0c}\vec{\mu}_{0c}\vec{m}_{a0}\} \tag{70}$$

$$R_{0a}^{(2)} = \sum_c \left[\frac{2\langle A_0 A_a | V | C_0 C_c \rangle}{(E_{0c}^2 - E_{0a}^2)}\right]^2 E_{0a} E_{0c} R_{0c}, \tag{71}$$

where R_{0c} is the rotational strength of the transition between the states of $|C_0\rangle$ and $|C_c\rangle$ in the chiral molecule.

Craig et al. [49] discussed the derivation of Eqs. (70) and (71) and they found that the two terms in the square brackets on the right-hand side of Eq. (70) are augmented for all achiral systems. The third term in Eq. (70) is singular to achiral molecules having magnetically-allowed transition(s) $\vec{m}_{a0}$, and then this contribution permits the direct assignment of a magnetic dipole transition in the achiral molecule.

Referred to as the induced optical activity, the resultant CD spectrum has measurable peaks only when the transition is either electrically or magnetically allowed. Thus, this ICD method is useful for discriminating the two types of transition [162].

Next, we consider ICD for random mutual orientation of an achiral and a chiral molecule. In this case, the first-order ICD [Eq. (70)] vanishes, since the angular factor G_{AC} averages to zero. The second-order rotational strength remains non-zero, because the square of the angular factor converges to 2/3. The first-order rotational strength can have both sings, plus and minus, depending on the mutual orientation of the chiral and the achiral molecules, whereas the second-order rotational strength is either plus or minus, since the right-hand side of Eq. (71) includes one square-bracket term only. Also, the first-order ICD depends on $1/R^3_{AC}$, whereas the second-order term ICD decreases abruptly, depending on $1/R^6_{AC}$. These characteristics of ICD are of importance for the following analyses;

i   discrimination of magnetic or electric dipole-allowed transitions in an achiral molecule [162],
ii differentiation of the mutual orientation modes, being random or fixed, between an achiral and chiral molecule [162],
iii determination of polarization direction of the transitions in an achiral molecule [162].

The characteristics are described in detail for many examples in the following.

3.4.1 Dispersion Force-Induced Circular Dichroism (DICD)

Axelrod[6], Bolard[7,8] and Hayward[9,10] have measured the ICD bands for the $n \to \pi^*$ transitions of carbonyl and nitro compounds in such solvents as CCl_4 and CH_3CN, with optically active substances. Axelrod[6] and Hayward[9,10] concluded that specific interactions, as those mentioned above, are not a prerequisite for DICD, and that these DICD phenomena may be observed in a system in which chiral and achiral species are assumed to be completely dissociated and interact through a long-range interaction such as van der Waals or dispersion forces. Schipper and Nordén[11] have discussed DICD for magnetic dipole-allowed transitions of achiral molecules through an interaction of the transitions in achiral molecules with the electric dipole transition(s) in chiral ones. The DICD has been mainly measured for $n \to \pi^*$ and d–d transitions which are electrically forbidden but magnetically allowed. In general, DICD is very weak, and occasionally single in sign. The DICD magnitude arising from an $n \to \pi^*$ is larger than that from $\pi \to \pi^*$. Also, the DICD decreases abruptly, depending on the fourth power of the energy difference of the given magnetic dipole transition in an achiral species and the electric dipole transition(s) in a chiral species [see Eq. (71)]. Thus, the inducing mechanism in DICD has been reasonably elucidated through Eq. (71), because a type of random orientation can be expected for the couple of an achiral species and a chiral species. Figures 12 and 13 show the absorption and CD of pyrazine and tetraphenylporphyrin in (+)-diethyltartrate, respectively. For pyrazine in (+)-diethyltartrate, the 310 nm band ($n \to \pi^*$) exhibits a larger CD than the 260 nm band ($\pi \to \pi^*$). The Soret band and Q band split into two components, x- and y-polarization, respectively, in the DICD for tetraphenylporphyrin perturbed through (+)-diethyltartrate as a solvent. Recently, Tajiri and Hatano[50] reported the $\pi \to \pi^*$ transitions of azulene to show CD in the presence of (+)-norpinone in solution. This suggests a weak interaction between azulene and the carbonyl moiety in (+)-

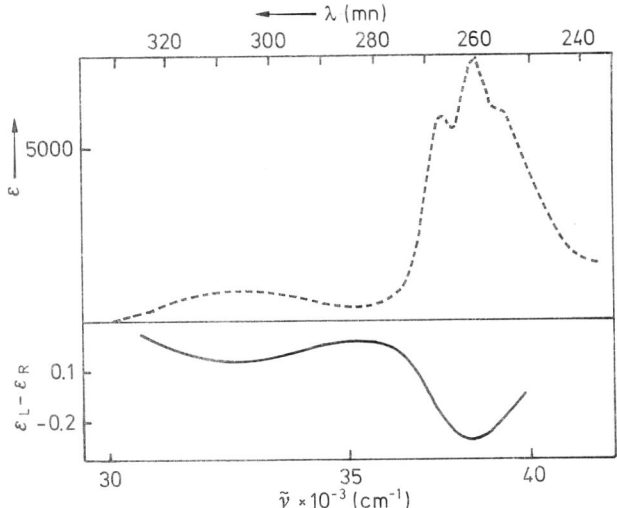

Fig. 12. Absorption (....) and circular dichroism (———) of 1,4-pyrazine in (+)-diethyltartrate [48]

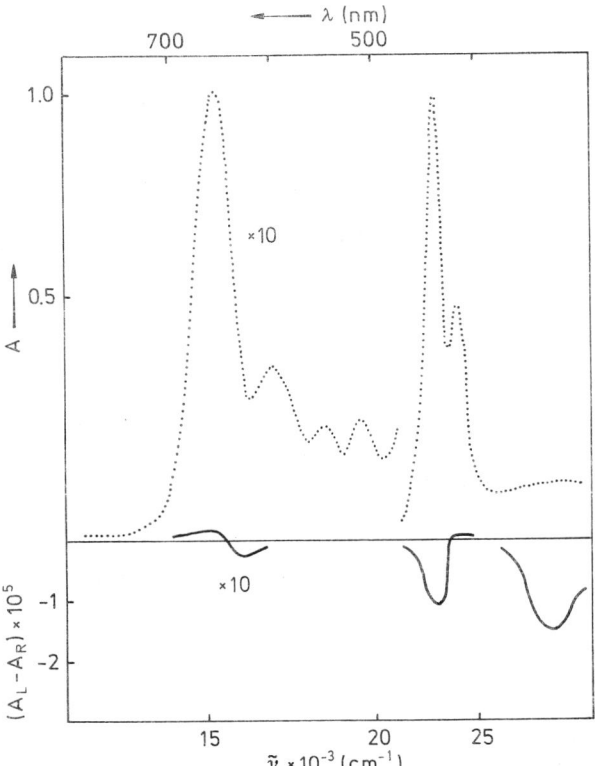

Fig. 13. Absorption (....) and circular dichroism (———) of tetraphenylporphyrin in (+)-diethyltartrate [48]

norpinone. The DICD technique is very useful for the analysis of interaction modes in the given molecular systems in solution.

3.4.2 Hydrogen Bonding-Induced Circular Dichroism (HBICD)

Achiral ketones [4,5,6,9], nitro-compounds [9], azo-compounds [9], thioketones [6], and nitrites [6] show ICD at wavelengths corresponding to their $n \rightarrow \pi^*$ transitions in the molecules when dissolved in a chiral solvent such as (—)-menthol. The ICD is of the order of 0.005 in $\Delta\varepsilon$. One can use chiral carbohydrates such as isosorbide as a chiral perturber. The magnitude of the ICD depends on the steric hindrance at the carbonyl group and also on its proton-accepting ability. Among the different aliphatic and alicyclic ketones which can form a type of hydrogen bond with a given chiral alcohol, acetone exhibits the largest ICD. Furthermore, the ICD magnitude also depends on physical conditions such as temperature or pressure. The inducing mechanism is not clear, but the HBICD may provide a powerful technique to analyze a cluster of solvent molecules around a solute molecule. On the other hand, the CD spectrum of chiral species changes often in magnitude and in sign when the hydrogen bond stabilizes a conformer which is otherwise unfavored. Thus, one needs to confirm the solvent effect [163] or temperature effect [164] on the CD, especially by using the octant rule projection. Thus, (M)-(+)-C_3-cyclotrignaiacyclene (*18*) and (S)-N-(1-methylpropyl)-β-cyclocitrylidene amine (*19*) exhibit a prominent temperature dependence on their absorption and CD when the temperature is lowered. For compound *19*, the results are summarized in Table 3.

M·(+)·*18* R = OCH$_3$
R'= OH

19

Table 3. Solvent and temperature dependence of the absorption (241 nm) and CD spectra of *19*

T, °C	$\varepsilon/1\ mol^{-1}\ cm^{-1}$		10^{40} R, c.g.s.	
	Isopentane	Methanol	Isopentane	Methanol
10	13700	5150	−0.58	−3.9
−20	13000	5400	−1.3	−4.6
−40	12300	5200	−1.6	−6.2
−60	11400	5000	−2.0	−7.2
−80	9300	4400	−4.2	−9.0
−100	7100	4300[a]	−8.2	−10.6[a]
−120	6600		−12.0	−12.6[a]
−140	6200		−16.6	−14.6[a]
−150	5900		−18.5	
−160				−17.0[a]

[a] Methanol/ethanol (1:4)

In methanol, the extinction of the absorption at 241 nm is less than 40% of the value in isopentane with little change on lowering the temperature. In isopentane, the extinction drops significantly to the value found in methanol when the temperature is lowered. The data on the temperature-dependent CD in methanol can be interpreted on the basis of a temperature-dependent equilibrium between two chiral species. The change of the rotatory strength appears to be $\Delta G° = -2.0$ KJ mol^{-1}. This phenomenon is interpreted by the assumption that the cis-conformer about the C(6)-C(7) bond is favored in methanol and, at lower temperatures, in isopentane.

3.4.3 Ionic Coupling-Induced Circular Dichroism (ICICD)

Complexes between chiral polymers having ionizable groups, and achiral small molecules become, under certain conditions, optically active for the absorption regions of the achiral small molecules. Dyes such as acridine orange and methyl orange have been used as achiral species, since they are in rapport with biopolymers through ionic coupling. This phenomenon has been applied to the detection of the helix chirality in poly-α-amino acids, polynucleotides, or polysaccharides when instrumental limitations prevent direct detection of the helices.

Occasionally the ionic coupling induces conformational changes around the chiral moiety, or an achiral species becomes optically active with the ionic coupling to form a favorable conformation. These perturbations of ionic species to the optical activity of chiral species can be examined by the experimental variation of adding ionic chemical species.

Experimental precautions will be discussed in Sections 4 to 6.

3.4.4 Ligation-Induced Circular Dichroism (LICD)

Bosnich [2, 3] has reported that when the complex Na$_2$[PtCl$_4$] is dissolved in (—)-2,3-butanediol, a CD band appears at only one magnetic dipole-allowed d–d transition and is directed to the z-axis which suggests that dissymmetric perturbations of optically active solvents are concentrated and directed to the tetragonal positions of the complex. Murakami and Hatano [51] have measured the ICD of achiral Cu(II)-β-diketonates dissolved in optically active nitrogen bases (D-α-phenylethylamine and nicotine). Figures 14 to 18 show the absorption and CD spectra of the five complexes: bis(acetylacetonato)copper(II), bis(benzoylacetonato)copper(II), bis(dibenzoylmethanato)copper(II), bis(dipivaloylmethanato)copper(II), and bis(diisobutyroylmethanato)copper(II), which are, respectively, abbreviated as Cu(acac)$_2$, Cu(bzac)$_2$, Cu(dbm)$_2$, Cu(dpm)$_2$, and Cu(dibm)$_2$, dissolved in d-(or l)α-phenylethylamine or nicotine. In addition, these figures involve the absorption curves in non-coordinating solvents. The present five Cu(II)-β-diketonates exhibited the same spectral behavior in α-phenylethylamine or nicotine as in pyridine. Therefore, it seems reasonable to expect that similar monoadducts are formed in the bases. The three complexes Cu(acac)$_2$, Cu(bzac)$_2$, and Cu(dbm)$_2$, in α-phenylethylamine exhibited similar CD spectra with one CD peak at the lower wave-number side in the ligand field region, while Cu(dpm)$_2$ and Cu(dibm)$_2$ showed different CD. The main CD of Cu(dpm)$_2$ is opposite in sign to those of the other four. In general, the CD in the ligand field region has been elucidated in terms of the mixing of the magnetic dipole transitions (in metal ions) with the electric dipole transitions (in ligand(s)). And it has been proposed that the

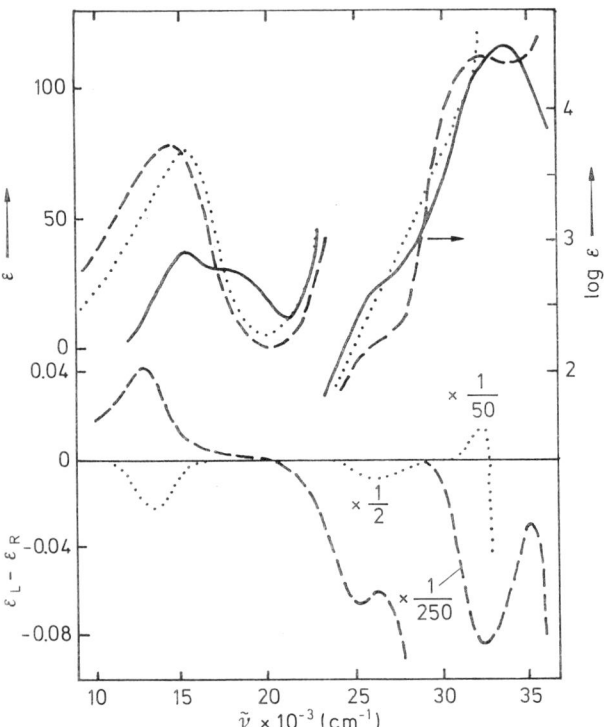

Fig. 14. Absorption and circular dichroism spectra of bis(acetylacetonato)copper(II) [51],
———, in chloroform;
– – –, in d-α-phenylethylamine;
······, in nicotine

order of orbitals of Cu(II)-β-diketonates in coordinating solvents is $d_{xy} \gg d_{z^2} > d_{x^2+y^2} > d_{xz} \approx d_{yz}$ if the effective molecular symmetry around the metal ion is taken as D_{2h}. The application of the order of d-orbitals to the five Cu(II) complexes leads to the assumption that the main CD band corresponds to $d_{z^2} \to d_{xy}$, since the main CD band appeared at the lower wave-number side within the d–d transition.

The three complexes Cu(acac)$_2$, Cu(dpm)$_2$, and Cu(dibm)$_2$, in the optically active solvents exhibited one or two CD bands in the vicinity of the absorption shoulder at about 25000 cm^{-1}. These bands are assigned to the charge-transfer transition between the metal d-orbital and the ligand π-orbital. Further, the intense CD band at about 30000 cm^{-1} is the π–π* transition of β-diketonate ions which is polarized along the long axis of the ligands. The transition shift to a lower wave number in the order of acac$^-$, bzac$^-$, dbm$^-$. Since the β-diketonates are optically inactive, the observed CD is assumed to be one kind of ICD, which is arising from the ligation of optically active amine to the Cu(II) complexes. The π → π* transition mixes with the transition of the chromophore in the optically active amine. According to the mechanism, the ICD is expected to decrease with the increase of the energy difference between the transition of the amine to be mixed with the π → π* transition and the π → π* transition itself. This expectation was confirmed by experimental results, as shown in Fig. 18.

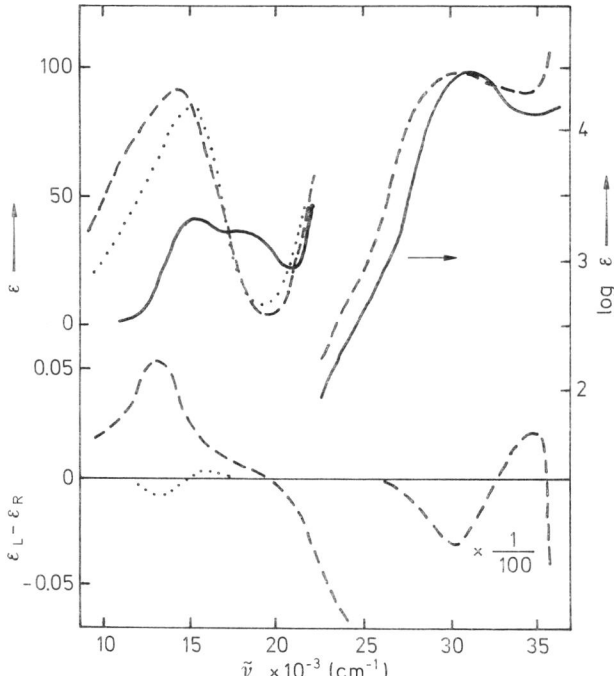

Fig. 15. Absorption and circular dichroism spectra of bis(benzoylacetonato)copper(II) [51],
———, in chloroform;
– – –, in d-α-phenylethylamine;
······, in nicotine

Thus, the CD magnitudes decrease with the reciprocal square of the energy difference, $1/(E^2_{amine} - E^2_{ligand})$, which is taken from Eq. (70).

Schipper [52] discussed the LICD in terms of group theory using Murakami and Hatano's experiments [51]. This new spectroscopic method permits the direct observation or discrimination of electric or magnetic dipole-allowed transitions.

3.4.5 Charge-Transfer-Induced Circular Dichroism (CTICD)

It is well known that intermolecular interactions between two organic compounds or organic and inorganic compounds give rise to a new absorption band(s), not present in the components. The new absorption band is characteristic of a loose reversible complex between an electron donor D and electron acceptor A, and ascribed to an electronic transition from D to A. The transition is called charge-transfer (CT) transition [53].

If either D or A is optically active, the CT transition is expected to exhibit CD, because the optically active species perturbe the CT transition. Firstly, Briegleb [54] observed the CD of the electron donor-acceptor (EDA) complex of various kinds of optically active ketones with tetracyanoethylene (TCNE). He concluded that the CT and CD are attributable to the $n \to \pi^*$ transition, where n is the lone-pair orbital of the optically

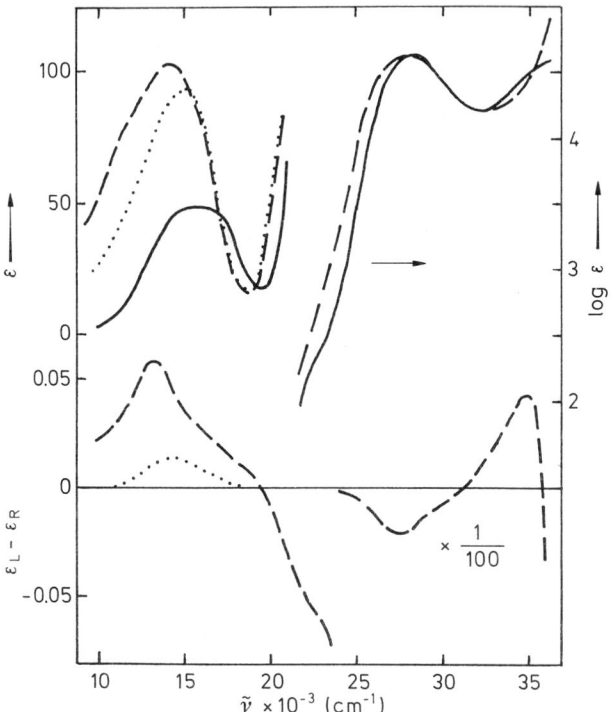

Fig. 16. Absorption and circular dichroism spectra of bis(dibenzoylmethanato)copper(II) [51],
——— ; in dioxane;
– – – ; in d-α-phenylethylamine;
...... ; in nicotine

active ketones and π* is the lowest unoccupied molecular orbital (LUMO) of TCNE. Further, Kuball et al. suggested that isomeric 1:1 complexes are present in the solution [55].

Tajiri and Hatano [56] have proposed a possible geometry of the complexes in solution. Thus, equilibria in EDA complexes of TCNE as an A and with (+)-fenchone, (−)-fenchone, (+)-camphore, (−)-camphore, and (−)-menthone as a D were discussed on the basis of absorption and CD measurements. Temperature dependence of solvent effects on the CT band also was investigated in addition to the measurements of magnetic circular dichroism (MCD) in solution to obtain more detailed information on the geometry and electronic structures of the complexes. Table 4 summarizes the thermodynamic and spectroscopic data for the CI complexes of the optically active ketones with TCNE. It is noteworthy that the (−)-menthone-TCNE complex exhibits a positive CD in polar solvents, while in nonpolar solvents the complex exhibits a negative CD and its magnitude is larger than that of the positive CD by a factor of 2. A conformational change in (−)-menthone itself can be assumed. Wellman et al. [57] have observed two maxima in the $n \to \pi^*$ region of its CD spectrum, as discussed in Sect. 3.4.2. A small positive CD at 300 nm increases in magnitude and a large negative CD at 285 nm decreases on changing the solvents from nonpolar to

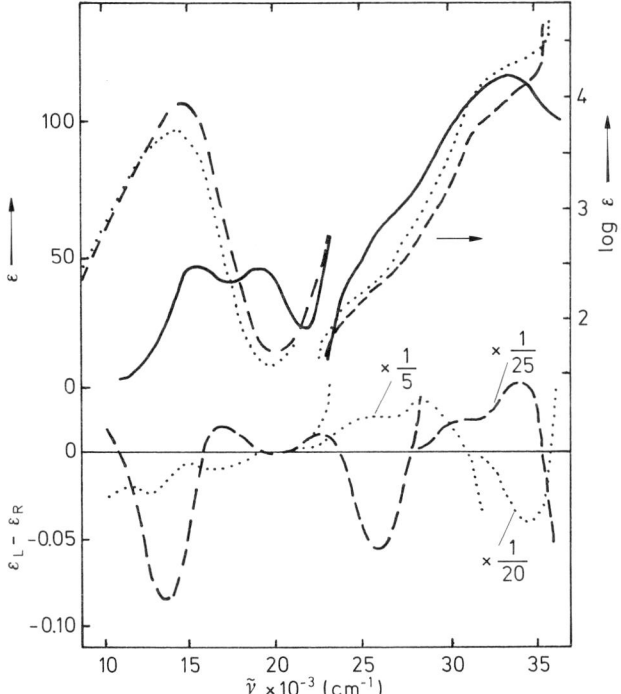

Fig. 17. Absorption and circular dichroism spectra of bis(dibenzoylmethanato)copper(II) and bis-(dipivaloylmethanato)copper(II) [51],
——, bis(dipivaloylmethanato)copper(II) in cyclohexane;
– – –, bis(dipivaloylmethanato)copper(II) in d-α-phenylethylamine;
......, bis(dibenzoylmethanato)copper(II) in l-α-phenylethylamine.

polar. The positive CD was assigned to the $n \to \pi^*$ transition occurring in a diequatorial conformer, and the negative CD to that in a di-axial conformer. Thus, (—)-menthone in solution is suggested to exist in an equilibrium, diequatorial ⇌ diaxial, although the equilibrium is markedly shifted toward the diequatorial conformer. For the fenchone- and camphore-TCNE complexes, the association constants from absorption, K_C^{AB}, are always larger than those from the CD, K_C^{CD}. In the menthone-TCNE complex, K_C^{AB} is larger than K_C^{CD} in all solvents except chloroform. The difference of K_C^{AB} and K_C^{CD} suggests the presence of isomeric complexes in the solutions. The possibility of isomeric complexes was first proposed by Orgel and Mulliken [58]. Unfortunately, no evidence for isomeric complexing has been obtained from the absorption and CD experiments. Based on the CD, MCD, and MO considerations, Tajiri and Hatano [56] proposed geometrical models for the complexes of optically active ketones with TCNE in solutions. The direction of rotation of TCNE around the carbonyl axis is of great importance. If the rotation is *clockwise*, an electron will transfer, on excitation, from D to A along the positive direction of the z-axis with *right helical motion*, i.e., the model assumes a left-handed helix (see Fig. 19). The above situation is called M-helicity. If the helicity rule is applied to the considerations of the given complexes, the model exhibits M- and P-helicity according to the *clock-*

wise and *counterclockwise rotation* of the TCNE plane around the z-axis of the carbonyl group. *M*-helicity exhibits negative CD. In the (+)-camphore-TCNE complex, the TCNE plane is expected to rotate *clockwise* around the z-axis so as to minimize the steric interaction with the methyl substituent on the C_1 atom of the ketone. An electron transfers along a left-handed helix upon excitation. The sign of the CD for the complex is minus. The model belongs to the *M*-helicity. In the (−)-fenchone-TCNE complex, the steric repulsion between the cyano group and the methyl substituents on the C_1 and C_3 atoms compels to get *M*-helicity (see Fig. 19). The TCNE complex with (+)-fenchone has the reversed helicity. This explanation is in accordance with

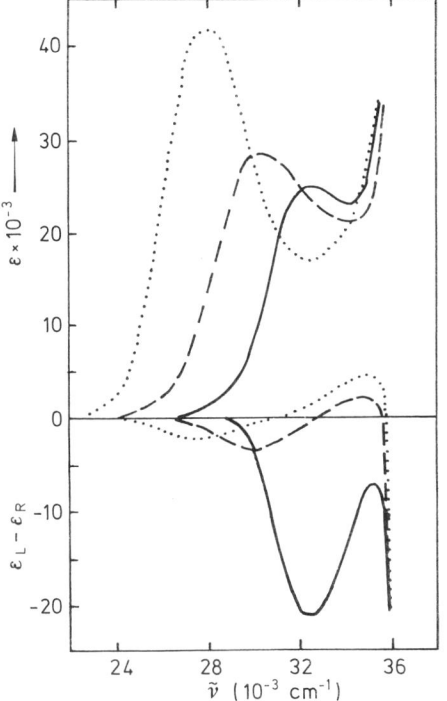

Fig. 18. Absorption and Circular dichroism spectra of the $\pi \rightarrow \pi^*$ band in *d*-α-phenylethylamine [51];
——, bis(acetylacetonato)copper(II);
– – –, bis(benzoylacetonato)copper(II);
....., bis(dibenzoylmethanato)copper(II)

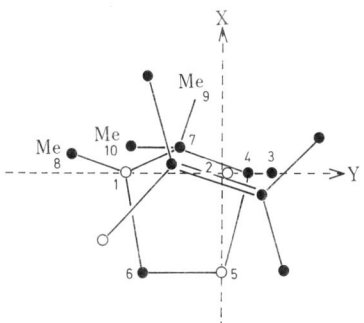

Fig. 19. Proposed geometry of the (−)-fenchone-TCNE complex [56]

Table 4. Thermodynamic and spectroscopic parameters for the charge-transfer complexes of the optically active ketones with tetracyanoethylene[56]

Donor	Solvent	Absorption		CD	
		$\varepsilon_{max}/\text{mol}^{-1} \text{ cm}^3 \text{ cm}^{-1}$	$K_C^{AB}/\text{mol}^{-1} \text{ dm}^3$	$[\theta]_{max} \cdot 10^{-3} \text{ deg mol}^{-1} \text{ dm}^3 \text{ cm}^{-1}$	$K_C^{CD}/\text{mol}^{-1} \text{ dm}^3$
(−)-Fenchone	CH_2Cl_2	1000	0.46	−0.65	0.39
	$CHCl_3$	522	1.45	−0.66	0.57
	CCl_4	473	17.28	−0.46	8.52
	C_6H_{12}	610 (571)	26.96 (31.2)	−0.56 (−0.99)	12.05
(+)-Fenchone	CH_2Cl_2	902	0.52	+0.60	0.51
(+)-Camphor	CH_2Cl_2	507	1.25	−1.14	0.45
	$CHCl_3$	726	1.14	−0.97	0.75
	CCl_4	783	8.33	−1.11	6.19
	C_6H_{12}	760 (774)	30.40 (29.6)	−1.18 (−1.32)	15.36
(−)-Camphor	CH_2Cl_2	536	1.16	+0.81	0.48
(−)-Menthone	CH_2Cl_2	643	0.76	+0.33	0.73
	$CHCl_3$	501	1.36	+0.13	3.50
	CCl_4	612	8.86	−1.16	7.70
	C_6H_{12}	667	21.24	−1.91	9.79

experimental results (see Table 4). The TCNE complex with (—)-menthone is expected to show *P*-helicity for the equatorial conformer in contrast to the *M*-helicity for the axial conformer.

3.4.6 Liquid Crystal-Induced Circular Dichroism (LCICD)

A chiral mesophase behaves as a chiral perturbation moiety for achiral molecules inserted to the mesophase. Mesophases are grouped into two classes based on the mechanism of the formation. These classes are thermotropic and lyotropic mesophases. The ICD in thermotropic cholesteric mesophases has been investigated by Saeva [59-62] and Sackmann [63,64]. They found that the signs of the ICD bands of achiral molecules which are intercalated in the mesophases depend on the sign and the wavelength of the cholesteric pitch band. The origin of the ICD is attributed to a type of helical arrangement of the achiral molecules in the mesophases. The cholesteric pitch can be estimated by [65]:

$$\lambda_{max} = nP, \qquad (72)$$

where λ_{max}, n, and P are the wavelength of maximum reflectivity, the average refractive index, and the cholesteric pitch, respectively. One can derive Eq. (72) by using the theory of light propagation in a cholesteric mesophase of de Vries [65] coupled with the

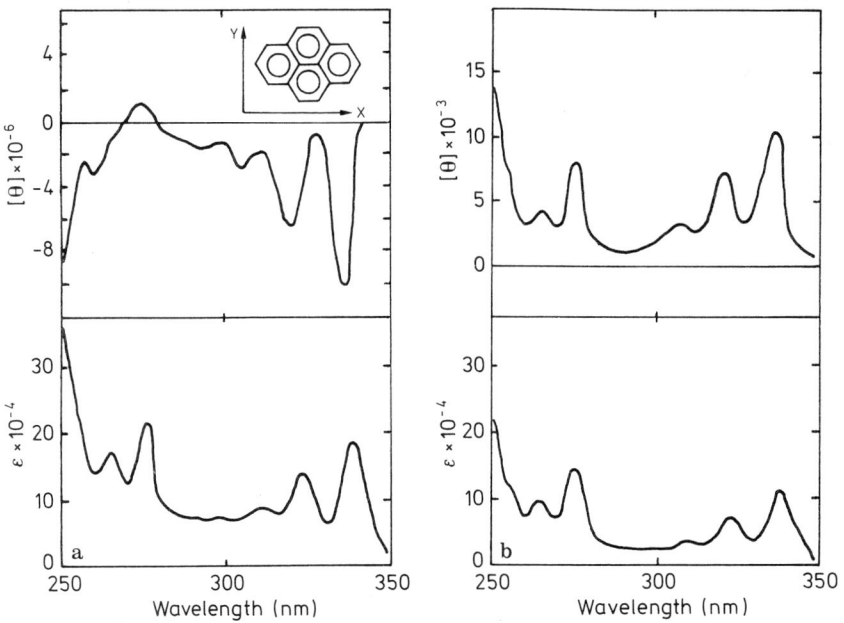

Fig. 20a and b. Circular dichroism and absorption spectra of pyrene in mesophases [67];
a. In thermotropic cholesteric mesophase, a mixture of cholesteryl nonanoate and cholesteryl chloride (weight ratio 7:3);
b. In twisted smectic mesophase of Tween 80-water system.
[pyrene] = 1 mmol l^{-1}

theory of circularly polarized light. Thus, the right-handed helix of a cholesteric domain exhibits a minus CD band as a pitch band, because right circularly polarized light is reflected selectively. This relation is occasionally inverted, especially in physical studies. Later Mason and Peacock [66] presented a more general equation applying to both isotropic and anisotropic solutes in cholesteric mesophases. This equation which is derived from Eqs. (70) and (71) explains the sign of the ICD bands of the solute molecules in a cholesteric mesophase, which correlates with the sign of the cholesteric pitch CD band. The sign depends on the polarization direction of the given achiral solute and the relative location in the wavelength of the respective ICD bands to that of the pitch band. This situation will be discussed later in detail.

Recently, Sato and Hatano [67-69] found a new type of chiral lyotropic mesophase composed of Tween 80, sorbitan mono-9-octadecenoate poly(oxy-1,2-ethanediyl), and water, and discussed the ICD of achiral solute molecules intercalated into the lyotropic mesophase. As the concentration of Tween 80 is increased, three distinct phases are obtained: micelle, neat phase, and reversed micelle, in that order. In the region of the volume ratio of Tween 80/(Tween 80 + water) of 0.40 to 0.63 under crossed Nicol-prisms, a focal conic texture was observed. This result indicates that the

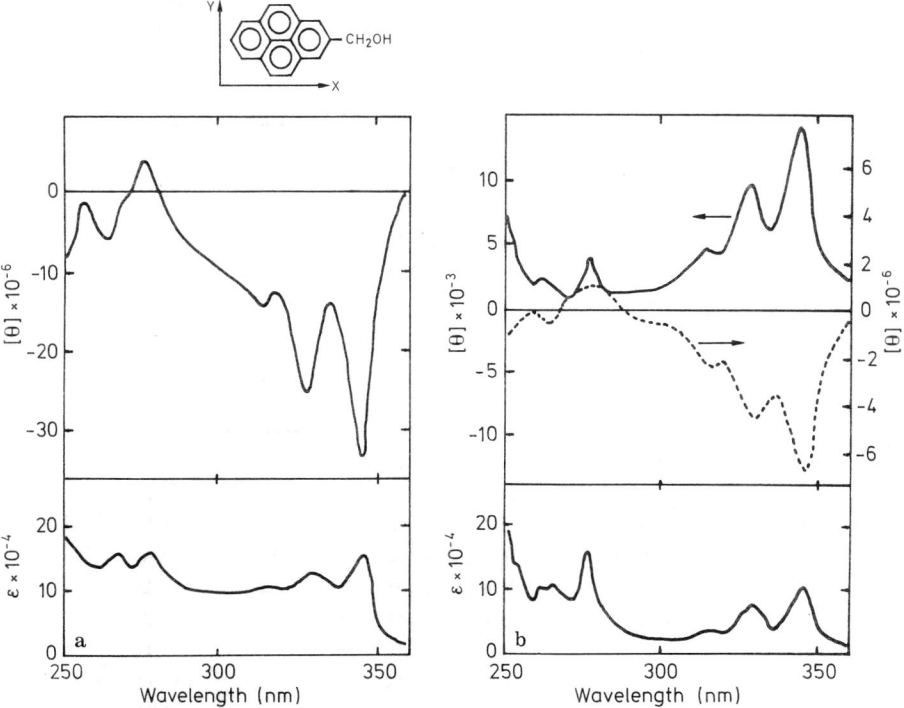

Fig. 21 a and b. Circular dichroism and absorption spectra of 4-hydroxymethylpyrene in mesophases [67];
a. In thermotropic cholesteric mesophase (constituents same as in Fig. 19).
b. In twisted smectic mesophase of Tween 80-water system.
[4-Hydroxymethylpyrene] = 1 mmol l^{-1} (———)
[4-Hydroxymethylpyrene] = 3.3 mmol l^{-1} (– – –)

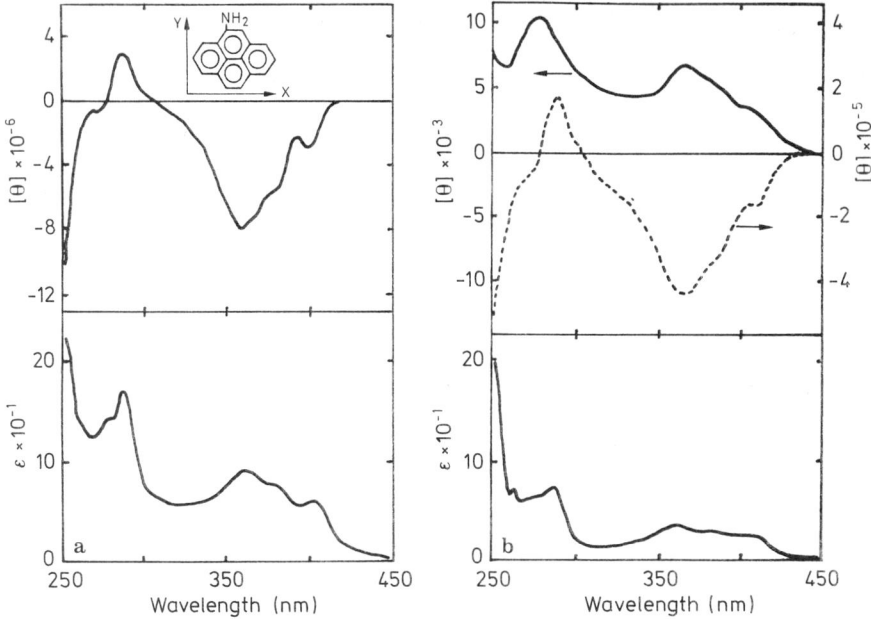

Fig. 22a and b. Circular dichroism and absorption spectra of 1-aminopyrene in mesophases [67];
a. In thermotropic cholesteric mesophase (constituents same as in Fig. 19).
b. In twisted smectic mesophase of Tween 80-water system.
[1-Aminopyrene] = 1 mmol l^{-1} (——);
[1-Aminopyrene] = 5 mmol l^{-1} (- - -)

neat phase, consisting of Tween 80 and water at equal volumes, is a lamella phase and truly a smectic phase. The smectic phase consists of the most highly ordered structure with molecules arranged in layers, with their long axes parallel to each other. The ICD and absorption spectra of pyrene, 4-hydroxymethylpyrene, and 1-aminopyrene in the thermotropic and lyotropic mesophases are shown in Figs. 20, 21, and 22. Although pyrene and its derivatives are achiral, the ICD of the derivatives was observed at wavelengths corresponding to their absorption when they were dissolved in the smectic phase of Tween 80-water. In the thermotropic cholesteric mesophase, the sign of the 0–0 band for the ^1B_b transition of pyrene at 277 nm is positive, whereas the sign of the ^1L_a band at 339 nm is negative. The former is polarized along the short-axis (y) and the latter along the long-axis (x). In the lyotropic smectic phase, these bands showed a single CD sign. However, when the concentrations of 4-hydroxymethylpyrene and 1-aminopyrene in the lyotropic phase were increased to about five times that of the above cases (see Fig. 21), CD spectra similar to those in the thermotropic phase were observed. This may result from an increase in the ordering of the solute molecules in the lyotropic mesophase. For pyrene, such a change in the CD spectra was not observed though the concentration of purene was increased to more than ten times the original. The difference in the spectral change depending on the concentration may be ascribed to the difference among their long- and short-axes polarization characteristics. The arrangement of the transition vectors of the long and short axes of the solute molecules can be regarded as *A* and *B* representations in C_2-symmetry, respectively, if the solute molecules in the mesophase preferentially orient their long molecular axis parallel to the alignment of the liquid crystal molecule. Thus, the principal axis of the solute molecule has a statistically preferred orientation relative to the direction of the local smectic layers. From Eqs. (70) and (71), one can easily understand that the sign of the ICD bands polarized parallel to the long axis of the solute molecules is the same as that of the CD band resulting from the pitch of the chirally twisted smectic mesophase if the pitch band is situated at a longer wavelength than those of the absorption of the solutes. The feasibility of the experimental determination of the CD does not only depend on the magnitude of the CD itself, but also on the ratio of the CD to the absorption coefficient, i.e., the anisotropy factor $g = (\varepsilon_l - \varepsilon_r)/\varepsilon$. The order of magnitude of g in the lyotropic mesophase is 10^{-4}, whereas it is 10^{-2} in the thermotropic mesophase. This may be due to the fact that in the thermotropic mesophase the degree of alignment of the solute molecules is considerably higher than in the lyotropic mesophase. This LCICD technique is very useful for the determination of the polarization direction of the given achiral molecule. Especially the Tween 80-water mesophase is preferable for spectroscopic applications, since it has the following requirements:

i   the solvent is a type of mesophase at room temperature;
ii   the solvent is transparent in the visible wavelength region;
iii no binding occurs between the solvent and added solute molecules.

3.4.7 Hydrophobic Interaction-Induced Circular Dichroism (HIICD)

Lately, ICD has been widely used for the analysis of the mode of interaction between a protein and a given substrate, also to examine the stoichiometry of the substrate bound to a protein, and the binding constants of the substrate. For example, it is well known that liver microsomes contain multiple molecular forms of cytochrome P-450. Binding of various hydrocarbons to P-448$_1$ (the major cytochrome component in liver microsomes from methylcholanthrene-treated rabbits) induces characteristic CD bands in the wavelength region where the bound hydrocarbons themselves show absorption but no CD band [158] (see Fig. 58). These bindings of substrates to enzymes result from the hydrophobic interaction of the substrate with the hydrophobic domain(s) in the enzyme. The hydrophobic interaction necessitates enzymes to be situated in an aqueous medium. The observed CD signs and magnitudes are depending on the fixness of the mutual orientation of the bound substrate with the chiral domain in a biomolecule, according to Eqs. (70) and (71). If the binding mode is tight and keeps a well-defined mutual orientation between the substrate and the biomolecule, the observed CD signs can be both plus or minus, and the CD magnitudes are to a great extent larger than the cases of loose binding modes. The intercalation of aromatic compounds into β-cyclodextrin (β-CDx) in aqueous media provides the best model for the former cases [162].

The CD signs of the ICD of β-CDx complexes with mono-, di-, and triazanaphthalenes in aqueous solutions depend on the polarization directions of these azanaphthalenes. The polarization assignments come from the CNDO/S-CI calculation [165] coupled with the MCD data of these azanaphthalenes [165]. In the first place, the origin of the coordinates was choosen in the center of a ring of β-CDx that is assumed to have a 7-fold symmetry axis. Next, $\vec{e}_i$, which is put at the origin of the β-CDx-fixed Cartesian coordinates, is expressed by the spherical polar coordinates (Fig. 23).

Assuming the symmetry considerations, we can obtain the final form of the rotational strength of the transition i from the ground state 0 to the excited state a in an aromatic compound R_{i0a}, expressed as:

$$R_{i0a} = \frac{1}{2}(3\cos^2\theta - 1)\mu_{i0a}^2 \times R_{zz} \,. \tag{73}$$

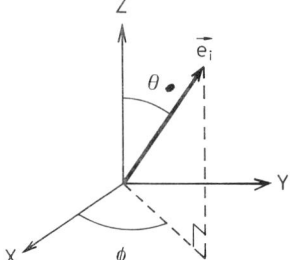

Fig. 23. Spherical polar coordinates for $\vec{e}_i$

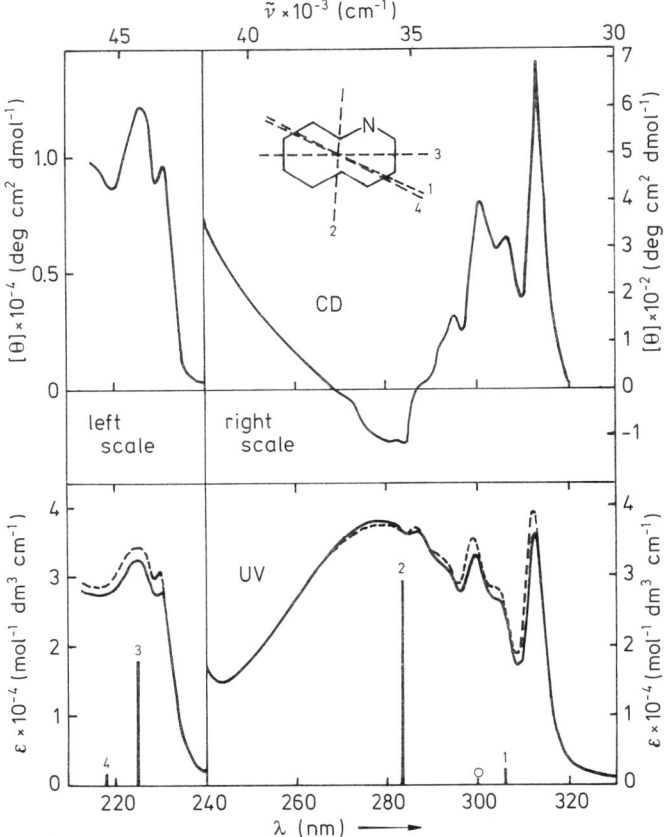

Fig. 24. CD (upper) and absorption (lower) spectra of β-CDx complex with quinoline in 1.0×10^{-5} M aqueous KOH (pH 7.45).
Broken line: in absence of β-CDx.
Solid line: β-CDx($1.438 \times 10^{-2}M$) + quinoline($3.226 \times 10^{-4}M$)

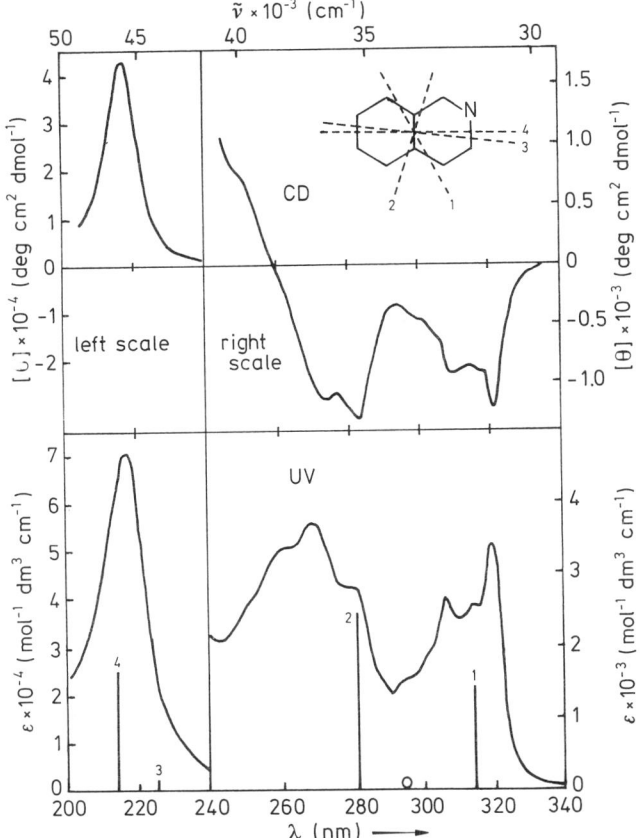

Fig. 25. CD (upper) and absorption (lower) spectra of the β-CDx complex with isoquinoline in 1.0×10^{-4} M aqueous KOH (pH 9.06): β-CDx(1.589×10^{-2} M) + isoquinoline(3.390×10^{-4} M)

Here, μ_{i0a} is the electric transition dipole moment of the transition i in a given aromatic molecule, and R_{zz} is the zz component of the inducibility on the rotational strength of the chiral species β-CDx. Since both μ_{i0a}^2 anh R_{zz} are positive, the sign of R_{i0a} depends only on the angle θ, which is the inclination angle of $\vec{e}_i$ with respect to the z-axis (the symmetry axis of β-CDx; see Fig. 23). Therefore, Eq. (73) indicates that the absolute direction of the transition dipole moment is determined by only the sign of the observed ICD spectra when the structure of intercalation complexes can be regarded as the axial intercalation in which the long axis of the guest molecule is parallel to the symmetry axis of the β-CDx cavity (z-axis). Figures 24 and 25 indicate the validity of the above-mentioned assumption. As shown in Fig. 25, the β-CDx complex with isoquinoline exhibited two negative ICD bands in the longer wavelength region, and these bands were attributed to the first and second π*←π transitions, which were both evaluated to be roughly short-axis polarizations. One intensive positive CD band derived from the third and fourth π*←π transitions was observed in the shorter wavelength region and coincided with the calculated results, in which both of the π*←π transitions were approximately long-axis polarized. For quinoxaline and cinnoline, weak negative CD bands were induced by the lowest $\pi^* \leftarrow n$ transition. Finally, it can be stressed that the ICD method using the intercalation phenomena of β-CDx is a useful tool for the analysis of the polarization direction of given electronic transitions in aromatic molecules [166]. More recently, Schipper and Rodger [167] have presented a symmetry rule for the analysis of the intercalation geometry of various β-CDx complexes using ICD.

4 Induced Circular Dichroism in Nucleic Acid-Dye Systems

The primary structure of nucleic acids consists of pyrimidine or purine bases linked via *N*-glucoside bonds to a *chiral* sugar residue, the sugars being linked by phosphodiester residues. Most cellular DNA consists of two covalently independent and complementary molecules arranged in an antiparallel helical double molecule. In the double helix, pyrimidine and purine bases are stacked so as the base pairs are spaced by 0.36 nm. The geometry of the DNA double helix was deduced from X-ray diffraction patterns. Analysis of such X-ray diffraction patterns reveals three different conformers of the DNA double helix, *A*-DNA, *B*-DNA, and *C*-DNA. The *A* and *B* forms are right-handed. *A*-DNA differs from *B*-DNA in that the base pairs of the latter overlap and that it is more compact. Spacing of the base pairs and inclination are the same; 0.36 nm and 20° to the planes perpendicular to the helix axis. The most striking difference between these two forms is the fact that the base pairs overlap in *B*-DNA making the structure so compact that, viewed from the top, the helix appears to be full of base pairs, while *A*-DNA has a void space around the helix axis. *C*-DNA is a distorted *B* form and its base pairs are somewhat more inclined to the plane parallel to the axis. Furthermore, single crystal analyses of short d(GC) oligomers have revealed a novel left-handed structure. This is called *Z*-DNA. Extensive studies of alternat-

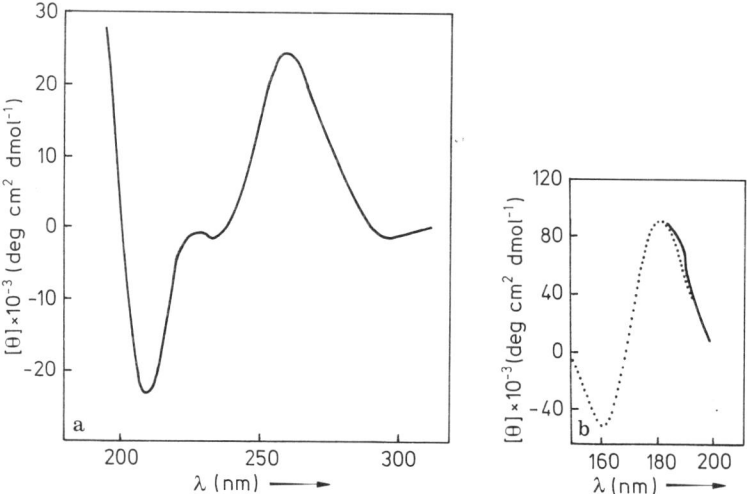

Fig. 26. Circular dichroism spectrum of double-stranded rice dwarf virus RNA in 0.15 *M* KF at 25 °C [86]. The relative CD magnitudes of the two negative CD bands at 210 nm and 240 nm are used as a discriminator for the A and B forms

ing d(GC) polymers and larger oligomers indicate that these molecules very probably form a similar left-handed Z-DNA structure in solution at high ionic strengths. Although the biological significance of Z-DNA is still uncertain, it is possible that the binding of specific proteins to GC-rich sites in DNA might effect a change of conformation from B to Z. The RNA double helix can only assume the A form, since it is impossible to accommodate the 2'-OH group of the ribose moieties in a B or C structure. Experimentally the A structure has been confirmed by X-ray diffraction studies in synthetic polymers such as pairs of poly I to poly C. Similarly, a DNA/RNA hybrid helix assumes the A conformation. In general, DNA assumes the B form in solution, and transforms its conformation from B to C when metal ions or amine is added to the solution.

4.1 Circular Dichroism Spectral Properties of Nucleic Acids

CD spectra provide the most conventional diagnosis for the conformations of DNA and RNA in solution. Figures 26, 27, and 28 indicate typical CD properties of A, B,

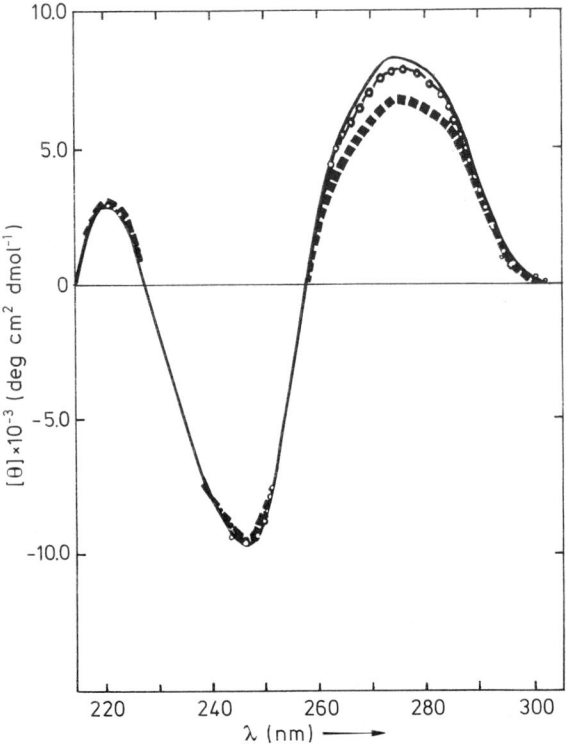

Fig. 27. Circular dichroism spectra of calf thymus DNA at pH 7 (27 °C) in dilute concentrations of electrolytes [72]. The spectra displayed above are the average obtained in 0.01 M and 0.04 M concentrations of NaCl (——), KCl (– – –), LiCl (OOO), CsCl (■●■●) and NH_4Cl (■■■). The molecular ellipticity of the negative band centered at 247 nm exhibited no significant change. However, the positive band at 275 nm changed in magnitude, when electrolyte is added. This shows that the B form is distorted toward the C form

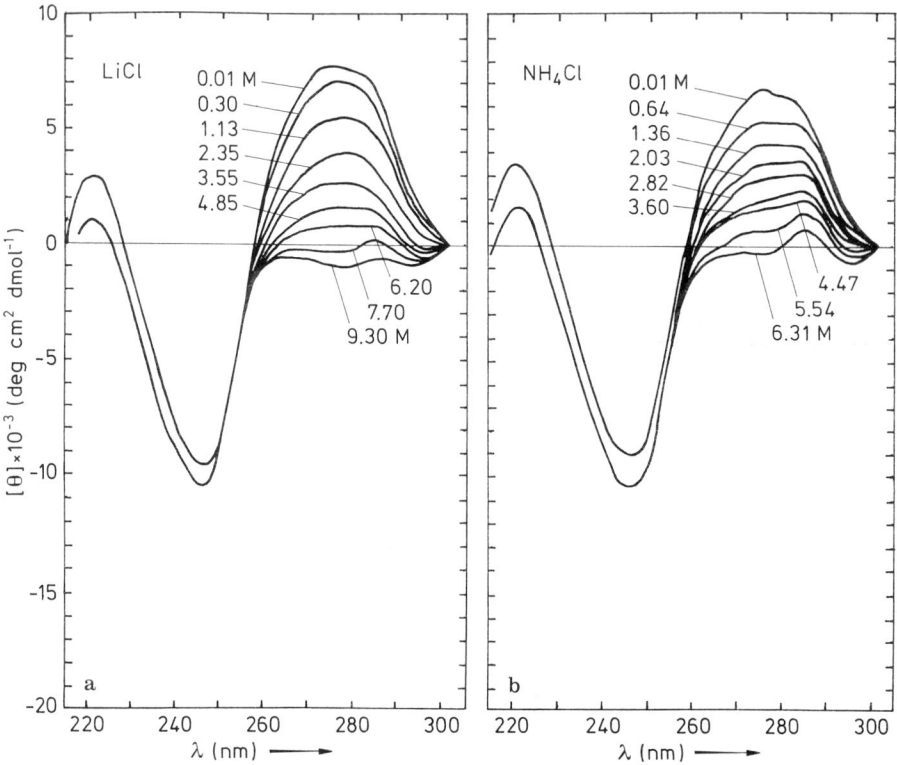

Fig. 28a and b. Circular dichroism spectra of calf thymus DNA at pH 7 (27 °C) in aqueous solutions of varying concentrations of LiCl (**a**) and NH$_4$Cl (**b**) [72]. The molar concentrations are printed above each spectrum. The decrease in the CD magnitude of the positive band at 275 nm demonstrates the transformation from the B to C form

and C conformers. The *A*-type conformer exhibits a tiny negative CD around 290 nm, a large positive CD in the region of 260–265 nm, and a large negative CD around 210 nm together with a shoulder at 235 nm. The *B*-form DNA exhibits a couple of large positive and negative CD bands in the region of 225–300 nm. The positive CD at 275 nm is markedly depressed, when the *B* form is converted to the *C* form by adding LiCl, NH$_4$Cl, or *n*-butylamine. Recently, Zimmerman and Pheiffer [70] examined the diffraction patterns of oriented fibers of salmon-sperm DNA immersed in some of the solvents which generate *C*-type CD spectra in solution and observed only *B*-form X-ray patterns up to concentrations of 9 *M* LiCl. The resolution of the controversy is difficult, since the transformed spectrum of protein-free DNA is only produced at high concentrations of salt or mixed water-alcohol solvents. Chen, Pheiffer, Zimmerman, and Hanlon have proposed to use *n*-butylamine as a denaturing reagent from *B* to *C*[71]. Thus, the positive CD band at 270 nm progressively decreases and its sign is inverted to minus as the concentration of *n*-butylamine in the solution is increased. These spectral changes in this wavelength region are very similar to those induced by increasing concentrations of simple electrolytes such as CsCl, LiCl, or NH$_4$Cl. The CD spectral changes induced by changes in concentrations of neutral electrolytes also have been discussed in detail by Hanlon et al. [72] As an example, the CD spectral change is depicted in Fig. 27. Much caution should be paid in CD observations of DNA in solution, since electrolytes or amine, if present in the solution, affect the CD spectral properties. X-ray crystallographical studies of the synthetic oligonucleotide d(CpGpCpGpCpG) demonstrated the left-handed, or Z-DNA, conformation [74, 75]. Earlier studies of fibers of the high-molecular-weight alternating

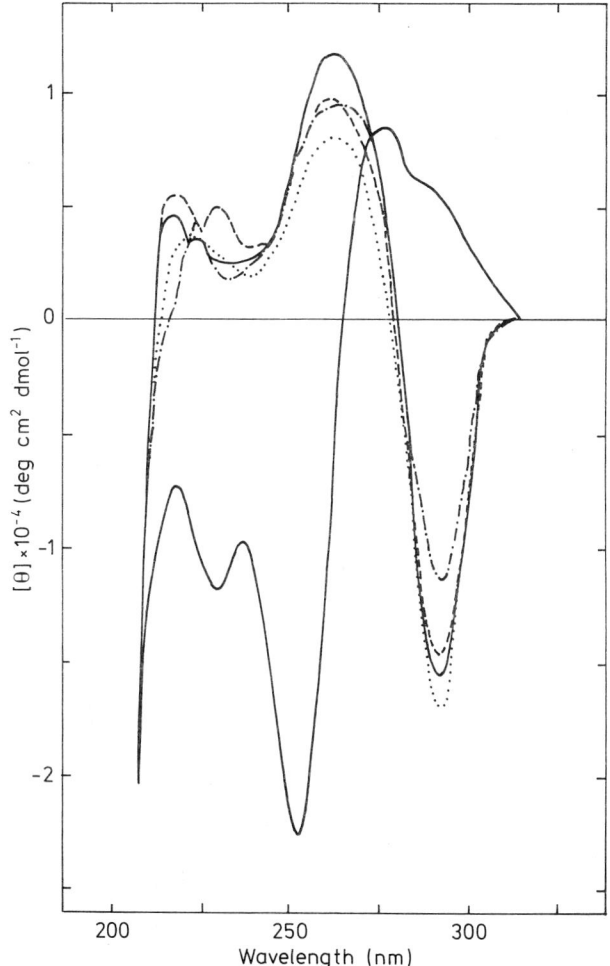

Fig. 29. Circular dichroism spectra of poly(dG-m^{5}dC) · poly(dG-m^{5}dC) in 5 mM Tris-HCl, pH 8.0, plus the following [73];
buffer only: ———;
1 mM BaCl$_2$: — — —;
0.02 mM hexaminecobalt(III): —·—·—;
5 mM NH$_3$(CH$_2$)$_4$NH$_3^{2+}$: ·······;
30% (vol/vol) ethanol: ——— .
The large negative trough at 290 nm corresponds to the formation of Z form. Co(NH$_3$)$_6^{3+}$, spermine^{4+} and spermidine^{3+} are the most effective for the B–Z transition

copolymer poly(dGdC) · ploy(dCdG) suggest that the Z form is also included in this copolymer [76]. These Z-form fibers can be obtained at high ionic strength and show unusual spectral properties. The sign of the CD spectrum of the Z form is inverted compared to that of the B form, and the ultraviolet absorption spectrum of the Z form shows an increase of the molar extinction coefficient compared to that of the B form in the range of 280–300 nm. A similar spectral behavior has been observed for another synthetic polynuceotide, poly[deoxyguanylyl-(3′-5′)-5-methyldeoxycytidine], poly(dG-m^{5}dC) [73]. These polymers have the B conformation at a very low ionic strength. The midpoint of the B–Z transition in poly(dGdC) occurs at 2.5 M NaCl, or 0.7 M MgCl$_2$. However, the transition midpoint in poly(dG-m^{5}dC) occurs at 0.7 M NaCl or 0.6 mM MgCl$_2$ (in the presence of 50 mM NaCl). A typical B–Z transition is shown in Fig. 29. The occurrence of such B–Z transitions has been highly interesting in analysing the induction mechanism of the off-switch of gene activity in DNA [77]. The B–Z transformation of double-stranded poly(dG-m^{5}dC) has been monitored by ^{31}P NMR spectroscopy as a function of the concentration of several neutral salts [78]. Infrared spectroscopy also has been utilized for the analysis of left-handed helical structures of poly[d(A-C)] · poly[d(G-T)] in the presence of Ni^{2+} ions [168]. Russell et al. have recently demonstrated that the very arginine-rich protein protamine stabilizes the Z form of poly[d(G-C)] [169]. Miller et al. [170] extended such observations and demonstrated that the nucleosomal core, containing both B and Z conformers of poly[d(G-C)], exhibits structural differences which may have important biological implications. Thus, a B to Z transition within a nucleosome in the (AC)$_n$-rich region influence the pack-aging of the protamine gene located several nucleosomes away, which is called "long-distance effect".

4.2 Feasibility of the Induced Circular Dichroism Technique for Nucleic Acid Research

Unlabelled nucleic acids are most commonly detected and measured by the following techniques: The first procedure depends on their absorbance at 260 nm. Double-stranded DNA at a concentration of 1 mg ml^{-1} has an absorbance of about 20, whereas a solution of a single-strand nucleic acid at the same concentration has an absorbance of about 26. The solution is monitored by CD spectroscopy, since each conformer of nucleic acids exhibits a characteristic CD spectral profile in the region of 250–300 nm. The second technique involes the staining of nucleic acids with dyes such as ethidium bromide or acridine orange, which fluoresce under ultraviolet irradiation after binding with nucleic acids. Nucleic acids are chiral. One can expect to detect ICD spectra after binding of the dyes to nucleic acids. The ICD arising from the binding of dyes provides very useful information for the analyses of conformational changes of nucleic acids and of the binding modes of the dyes to nucleic acids.

Neville and Bradley [79] were first to report that DNA induced Cotton effects in the optical rotatory dispersion spectrum of acridine orange. The investigation of DNA-induced Cotton effects has been extended to other aminoacridines. According to Peacock [80], aminoacridines may be divided into two groups. The DNA-induced ICD of the first class (aminoacridines having a 3-amino group) increases steeply and cooperatively with increasing amounts of bound aminoacridine r, where r is defined as the total bound dye concentration divided by the total DNA phosphate residue concentration. The second class are aminoacridines lacking a 3-amino group, which vary little with r. At low values of r, the ICD is weaker but directly due to the interaction of a single adduct with the DNA chromophore. The ICD is evidence for the binding of the dye to the DNA. Unbound dyes could in principle have a small DICD. When the concentration of the dye is increased, a part of the bound dye molecules is intercalated into the base pairs in DNA. The intercalated dyes may induce a conformational change of the DNA itself, and then the CD spectra must be separated from those arising from the DNA. Especially the CD in the region of 250–300 nm includes both origins of the DNA and added dye molecule bound to the chiral moiety of DNA. Visible region ICD is useful for the analysis of binding modes of dye molecules. Schipper, Nordén, and Tjerneld [81] proposed a theoretical model predicting the signs of the ICD of methyl orange, proflavin, and pseudocyanine bound to DNA. The signs indicate the degree of alignment of the bound dye molecules relative to a DNA axis. If the polarization direction of the given dye molecule is known, the binding mode can be predicted through the signs of the ICD.

However, the ICD technique is limited to know the specific nucleotide units or the base pairs in DNA interacting with a given dye molecule and to define the loci where the local units in the dye molecule and in the nucleotide moiety are interacting with each other, because the ICD provides the averaged structural information only. Therefore, the DNA-dye interactions have been studied on the basis of X-ray structural analysis of the drug-dinucleoside monophosphate complexes [82]. NMR spectroscopy is more useful for direct evidence of both their interaction mode and the involvement of hydrogen bonding in their selective recognition of certain base pairs [83].

Therefore, the ICD method should be applied to the analysis of the DNA-dye complexes together with other physicochemical structural analyses such as NMR spectroscopy or X-ray crystallographical analysis.

4.3 Conformational Changes of Deoxydinucleoside Monophosphate and Polydeoxyribonucleotides Induced by Added Drugs or Metal Ions

In Sect. 4.1 it was described that the left-handed Z form of double-strand poly(dG-m^{5}dC) · poly(dG-m^{5}dC) is formed at much lower salt concentrations than for the unmethylated copolymer. The B–Z transformation has been investigated with ^{31}P NMR [84], IR [168], and CD [85] spectroscopies. Adriamycin (Formula 20) is a cytotoxic

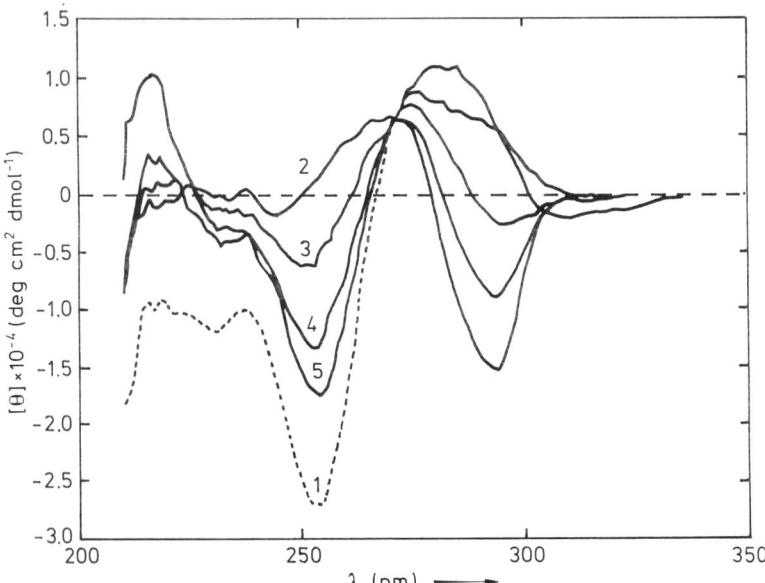

Fig. 30. CD spectra (molar ellipticity [θ]) at room temperature showing the effect of adriamycin on poly(dG-m⁵dC) · poly(dG-m⁵dC) conformation [85].
(1) Original B form of poly(dG-m⁵dC) in the presence of 5 mM Tris/50 mM NaCl/0.1 mM EDTA, pH 8.0.
(2) Z form after addition of $MgCl_2$ to (1) (final concentrations; 100 μM nucleotides, 2 mM $MgCl_2$).
(3–5) Addition of adriamycin to (2).
(3): 2.0 μM, [adriamycin]/[nucleotide] = 0.02 or 1:50.
(4): 4.0 μM, [adriamycin]/[nucleotide] = 0.04 or 1:25.
(5): 8.8 μM, [adriamycin]/[nucleotide] = 0.091 or 1:11

antibiotic drug, and is used widely in cancer chemotherapy. This drug binds strongly to DNA with which it forms an intercalation complex. Addition of adriamycin to poly(dG-m⁵dC) in the Z form resulted in a conversion to the B form (see Fig. 30).

Poly(dG-m⁵dC) · poly(dG-m⁵dC) appears as a B form double helix in the presence of 5 mM Tris/50 mM NaCl/0.1 mM EDTA, and the B form is converted to its Z form after addition of $MgCl_2$ (final concentrations: 100 μM nucleotides, 2 mM $MgCl_2$). But, adriamycin converted the Z form of poly(dG-m⁵dC) in the presence of $MgCl_2$ to the B form. Figure 30 shows the B → Z → B transition of the polymer. CD data provide the degree of cooperativity of the B to Z transition of poly(dG-m⁵dC) by using a modified Hill equation for an equilibrium between two states:

$$y = y_\infty + (y_0 - y_\infty)/[1 + (x/K)^n], \tag{74}$$

where y is the spectroscopic response, x the concentration of added reagent, y_0 the response when $x = 0$, y_∞ the maximal response when $x = \infty$, K the concentration of x at the midpoint (K is equivalent to the equilibrium constant), and n is the degree of cooperativity that determines the slope of the curve. From the CD data at two wavelengths (293 and 252 nm), the percent change was estimated and the concentration expressed as log x. These CD data as a function of the [adriamycin]/[nucleotide] ratio were fitted with Eq. (74), and it was found that adriamycin alone converts the Z form to the B form in a cooperative way with $n = 2$. This experiment indicates that adriamycin binds preferentially to the B form and converts the Z form to the B form. This drug exerts an influence on the active gene by activating inactive genes in a tumor.

Temperature-dependent B–Z transitions also have been reported [87]. Poly[d(A-m^{5}C) · d(G-T)] exhibits a dramatic inversion of the CD spectrum in sign when the temperature is raised from 46 to 68 °C. This change in the CD spectrum corresponds to the B–Z transition. A similar transition occurs when the concentration of neutral salt is raised. The helical form in high salt and at high temperature was designated as left-handed, and the corresponding low-salt or low-temperature form as right-handed. Analysis of the thermal transition in terms of the van't Hoff plots gives an effective enthalpy change of $+0.92$ kJ mol^{-1} and an entropy increase of $+2.7$ kJ K^{-1} mol^{-1}. The B–Z transition was investigated by ^{31}P NMR and laser Raman spectroscopies. Recently, together with the finding that a duplex of poly(dC-dG) adopts a left-handed form termed Z-DNA, syn-anti conformation change around the bond between the base and deoxyribose moiety has been discussed [88]. In poly(dC-dG) at high salt concentrations, guanine favors to form a syn-conformer. To investigate the syn-anti conformation change, deoxydinucleoside monophosphate has been used as a model compound for DNA [89]. Zamecnik [90] has discussed the conformation change of dinucleoside oligophosphates influenced by 0-methylation in ribose moiety and metal interaction in terms of CD.

Poly(8-methyladenylic acid) can form a regular, single-strand helix, the first such structure found to exist in aqueous solution. In this polymer, the glycosidic conformation is a regularly alternating syn-anti structure. It is interesting that the alternation of syn and anti conformations may be correlated with Z-DNA formation and that the bulky 8-methyl group is expelled to the outside of the stack's array of bases [171].

4.4 Metal Complex Binding to DNA

Recently, there has been increased attention focused on the binding of metal complexes to DNA, as the antitumor drug cis-[Pt(NH$_3$)$_2$Cl$_2$] binds to DNA [172]. Therefore, it is of interest to determine the stereochemistry of cis-[Pt(NH$_3$)$_2$Cl$_2$] binding to DNA. Spectroscopic studies of the binding of cis-[Pt(NH$_3$)$_2$Cl$_2$] to mono- and dinucleotides reveal the heterocyclic nitrogen atoms of the purine and pyrimidine bases to be most favorable DNA binding sites [173]. Lippard et al. showed that cis-[Pt(NH$_3$)$_2$Cl$_2$] binds to regions of DNA rich in (dG)$_n$ · (dC)$_n^2$ ($n \geq 2$) sequences [174]. Furthermore, the same authors have examined the reaction of cis-[Pt(NH$_3$)$_2$Cl$_2$] with the self-complementary deoxyribohexanucleoside pentaphosphate [d(ApGpGpCpCpT)]$_2$ for elucidating the mechanism of action of platinum antitumor drugs [175]. Divalent platinum binds covalently to the bases and the binding sites can be determined from plots of the pH dependence of the chemical shifts of the nonexchangeable base protons for both the unplatinated and platinated oligonucleotides. Upon binding of cis-[Pt(NH$_3$)$_2$Cl$_2$], all nonexchangeable base protons shift to lower magnetic fields. The behavior of the base protons is entirely analogous to that of the free oligonucleotide at lower pH. The observed shifts from pH 5 to 9 are assigned to the protonation of N-1 of guanine. The apparent pK_a (8.0) for the deprotonation of guanine N-1 is less than that for the free oligonucleotide (9.8). Such a decrease in the pK_a of the N-1 is indicative of the guanine complexes in which cis-$[Pt(NH_3)_2Cl_2]$ is bound at the N-7 position. The result, coupled with the characteristic $0.3 \sim 0.9$ ppm downfield shift of guanine H-8 observed upon Pt(II) binding to N-7, reveals that cis-$[Pt(NH_3)_2Cl_2]$ is ligated by the N-7 atoms of guanine residues. The significant downfield shifting upon platination of resonances belonging to bases not directly bound to platinum can be observed, and this suggests that a significant degree of base unstacking accompanies the binding of cis-$[Pt(NH_3)_2Cl_2]$. Kidani et al. [176] digested the adduct of (R,R)-1,2-cyclohexanediamine platinum(II) with DNA by endo- and exonucleases and the digested products were fractionated by means of HPLC. The absorption spectrum of the product showed an absorption maximum at 260 nm, and the molar extinction coefficient of platinum was 22 500 at 260 nm, indicating that the binding ratio of Pt to d(GpG) is 1:1. The product did not show any absorption change in the pH $1.0 \sim 6.5$ range, suggesting that the N-7 position of guanine had already been occupied by the Pt(II). Recently, Barton et al. [177,178] reported chiral discrimination in the covalent binding of bis(phenanthroline)dichlororuthenium(II) to B-DNA. In buffer containing 10% ethanol, 50 mM $NaNO_3$, 5 mM Tris at pH 7.1, racemic bis(phenanthroline)-dichlororuthenium(II) (50 μM) ($Ru(phen)_2Cl_2$) was incubated at 37 °C with DNA (500 μM nucleotide). Then, NaCl and 95% ethanol were added to quench the reaction. One Ru(II) complex binds to 11 base pairs for 3.5 hrs. Under the same conditions, the coordinatively saturated tris(phenanthroline)ruthenium ($Ru(phen)_3^+$) cation, which may be intercalated into DNA base pairs, did not bind to DNA. The super-

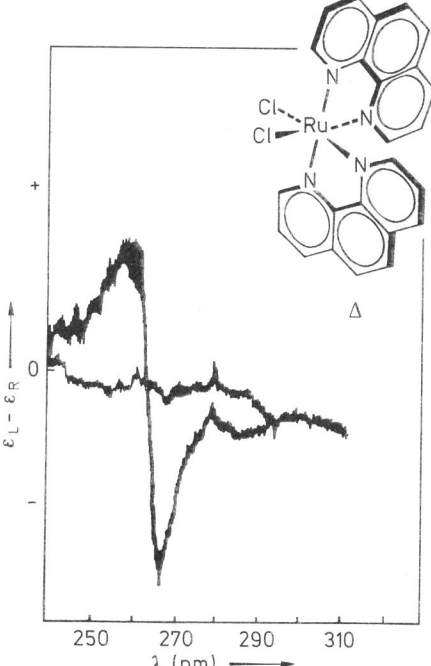

Fig. 31. CD of the supernatant after ethanol precipitation of the ruthenium complex bound B-DNA. Binding to B-DNA is stereoselective and leads enrichment of the supernatant in the unbound Δ-isomer [178]

natant from the incubated mixture of Ru(phen)$_2$Cl$_2$ with DNA exhibited a couple of CD bands centered at 262 nm, the CD magnitudes being 5 times larger than those obtained for intercalated Ru(phen)$_3^+$ into DNA. Hence the degree of chiral selectivity for this covalent adduct, Ru(phen)$_2$-DNA, appears substantially greater than for the Ru(phen)$_3^+$ cation.

Figure 31 shows the CD of the supernatant after precipitation of the Ru(phen)$_2$ bound to *B*-DNA. This CD indicates that the binding enriches the Δ-isomer (inset) of Ru(phen)Cl$_2$ in the supernatant containing the unbound ruthenium complex. Thus, Λ-Ru(phen) binds preferentially to *B*-DNA. The binding site would be the N-7 nitrogen atom of guanine, which is readily accessible in the major groove. On the contrary, the Δ isomer of the Ru(phen)$_3$ cation is intercalated with the *B*-DNA [179]. The Δ isomer of Ru(phen)$_3$ has the same helical screw sense as the right-handed DNA. On the other hand, the metalation of the bases seems to require the Λ-configuration.

The possible similarity between Ru(phen)$_2$Cl$_2$ and *cis*-[Pt(NH$_3$)$_2$Cl$_2$] in interactions with *B*-DNA is of much interests, since antitumor activities and toxicities of various ruthenium complexes have been recently reported.

4.5 Nucleic Acid-Protein Systems

Details of core-particle DNA secondary structure, which consists of DNA and the very lysine-rich histone H1, have not yet been determined, but Raman spectroscoppy [181] and wide-angle X-ray scattering data [182] indicate that it is closely similar, if not identical, with that of *B*-DNA. Cowman and Fasman utilize the CD variation of DNA for the analysis of DNA conformation changes upon the formation of nucleoprotein complexes from DNA and histone [183]. The CD spectra for core particles and H1- and H5-depleted mononucleosomes are compared to the CD spectrum of protein-free DNA. The CD spectra for the H1- and H5-depleted mononucleosomes exhibit both lowered magnitudes and altered band shapes, relative to DNA alone. The CD spectrum of nucleosomes between 250 and 320 nm can be resolved in two components. One corresponds to the secondary structure of protein-free DNA (*B*-DNA), and the other is a negative band, centered at 275 nm, which is attributed to the condensation of *B*-DNA into an asymmetric tertiary structure. The CD analysis leads to estimation of the base lengths which interact with the histones [184].

4.6 Future Trends and Scope on Induced Circular Dichroism in Nucleic Acid-Dye Systems

CD spectra provide successfully the most conventional diagnosis for the conformations of DNA and RNA in solution. Further, the interaction modes between a given drug and nucleic acid and the quantitative estimation of their binding or dissociation constants have been verified by CD observations. More detailed, atom-resolved information on the conformational changes in DNA can be obtained complementarily from other spectroscopic measurements such as NMR and Raman spectroscopies [185]. Interaction sites are clarified by NMR experiments, as mentioned above. The base stacking sites also can be analyzed in terms of each nitrogen atom in their bases by using super-high-field NMR observations. However, the CD technique can present averaged but quantitative information on the stacking modes and conformations of a given DNA and RNA. The most pressing topic in medical research today is cancer research, and here especially the field of oncogenes (cancer-inducing genes). These are genes that, when activated in cells, can transform the cell from normal to cancerous. A single-chain polypeptide with a molecular weight of 10,000 to 11,000 daltons has been sequenced which may behave as a tumor-inhibitory factor. The specific interactions between inhibitory peptides and oncogenes are of much interest in oncogene research. We emphasize the great importance of the precise CD measurement for samples containing genes and polypeptides at the very low concentrations.

5 Induced Circular Dichroism in Protein-Dye Systems

The circular dichroism (CD) technique has been widely and conveniently used for studying the conformations and conformational changes of proteins [90a, 91-96]. The CD spectrum of a protein reflects the overall picture of the molecule. It depends on all amide groups. However, it is well known that no single reference CD spectrum can accurately represent all members of proteins. Thus, conformations in proteins eventually deviate from the ideal model forms. The mean residue ellipticity depends on chain length and number of strands in the β sheet; the contribution of non-peptide chromophores can not be neglected. For particularly large proteins containing many domains, it would be more meaningful to investigate the tertiary structure for each domain. This may be done by dividing the proteins into major domains and subjecting each domain to CD analysis. Prediction of the entire structure of the proteins can be practiced by combining the analysis of CD spectra for the fractions of secondary structures with that of tertiary structures to aid in elucidating the structure of proteins that are not crystallized. For the past few years gene manipulation techniques have been used tried in an attempt of replacing a given amino acid residue in a protein by another in order to synthesize more thermally stable proteins and more effective enzymes than native ones. Prediction and precise analysis of the entire protein structure in solution are becoming indispensable in protein chemistry.

On the other hand, X-ray structural analyses of protein single crystals have been reported during the past decades. Crystallographical studies provide much information on the stereochemistry of the protein family. An alternative approach has been proposed to predict the secondary structures of given proteins in solution, using CD spectra of well-defined proteins crystallographically. Eight types of secondary structures are considered: α helix; parallel and antiparallel β sheet, types I, II, and III β turn, all other β turns, and other structures. This method is based on a mathematical calculation of orthogonally based CD spectra from the CD spectra of proteins with known secondary structure. If the primary structure, i.e., the amino acid sequence, is known, one can approximately predict a plausible secondary structure for the given protein in solution. The limitation and scopes of this method will be discussed in the followings.

In enzymology, binding constants of substrates or inhibitors of enzymes are estimated by registering the changes of CD the magnitudes induced at the wavelengths of the added molecules. This induced circular dichroism (ICD) depends on the binding mode and the chemical quantity of the bound molecules. This chemical measure is essential

5.1 Estimation of the Contents of Each Fraction and of the Secondary Structure

At first, estimations were carried out by using optical rotatory dispersion (ORD); this was necessary because of instrumental limitations. Conventionally, the Moffitt-Yang equation [Eq. (23)] can be modified to:

$$[m'] = \frac{(a_0^R + x^H a_0^H + x^\beta a_0^\beta)\lambda_0^2}{\lambda^2 - \lambda_0^2} + \frac{(b_0^R + x^H b_0^H + x^\beta b_0^\beta)\lambda_0^4}{(\lambda^2 - \lambda_0^2)^2}. \tag{75}$$

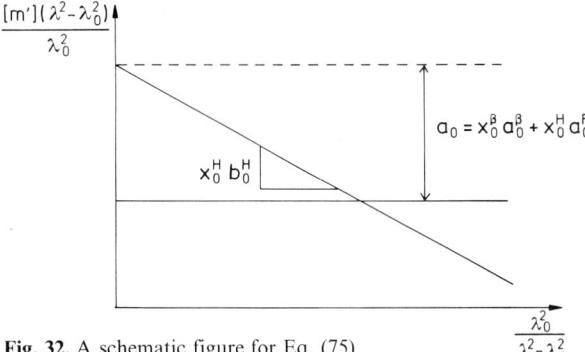

Fig. 32. A schematic figure for Eq. (75)

On plotting $[m'](\lambda^2 - \lambda_0^2)/\lambda_0^2$ against $\lambda_0^2/(\lambda^2 - \lambda_0^2)$, one gets $b_0^R + x^H b_0^H + x^\beta b_0^\beta$ and $a_0^R + x^H a_0^H + x^\beta a_0^\beta$ from the slope and intercept in the plots (see Fig. 32). If the values of a_0^R, a_0^H, a_0^β, b_0^R, b_0^H, and b_0^β are known, the contents (fractions) x^H and x^β, the α helix and β-structure are determined from the observed ORD curve. If $\lambda_0 = 212$ nm, $a_0^R = -580$, $a_0^H = +680$, $a_0^\beta = +800$, $b_0^R = 0$, $b_0^\beta = 0$, and $b_0^H = -630$ [97)98)].

Recent development of circular dichroism spectrometers provides more precise measurements (ellipticity or CD magnitude with its sign) for estimating the secondary structure in proteins. CD spectra in the far ultraviolet region include amide $n \to \pi^*$ transitions at 222 nm or below, aromatic amino acid residues are registered at 230 nm and 280 nm regions, and the disulfide chromophore between 245 and 280 nm. In addition, there can be contributions from prosthetic groups such as heme, complexed metal ions, and bound coenzymes. The ORD method is less useful for common proteins. The difficulty in interpreting the observed CD results in separating the amide contributions which depicts an average picture of the secondary structure from other contributions which are sensitive to the side-chain conformations. CD is more versatile and useful than ORD, because each CD contribution results from a CD origin which can be conventionally separated. The relationship between the secondary structure of proteins and the CD contribution of the amide chromophore has been extensively investigated in theory [99,100] and experiment [101,102]. For example, the CD spectra of ribonuclease A, S, S′, and S′ protein include a negative near-ultraviolet CD band at 275 nm and a positive CD band of varying intensity near 240 nm. The negative band at 275 nm is thought to arise from tyrosine residues with additional contributions from disulfides. Goux and Hooker, Jr. separated the side-chain contribution to the observed CD and assigned the CD origin theoretically [100]. Chen et al. [103] showed that the CD spectrum of a protein can be expressed in terms of Eq. (76), where $f_H + f_\beta + f_R = 1$ and f_H, f_β, and f_R are fractions of α helix, β sheet, and random-coiled structure, respectively.

$$[\theta]_\lambda = f_H[\theta]_\lambda^H + f_\beta[\theta]_\lambda^\beta + f_R[\theta]_\lambda^R \tag{76}$$

$[\theta]_\lambda$ is the experimental mean residue ellipticity at wavelength λ (in nm), and $[\theta]_\lambda^H$, $[\theta]_\lambda^\beta$, and $[\theta]_\lambda^R$ are the theoretically computed values from the X-ray studies of nine different proteins for 100% helix, 100% β sheet, and 100% random coil, respectively. Chen et al. proposed to use a chain length-dependent factor k, as $[\theta]_\lambda^H$ depends on the average number of amino acid residues per helix (n) leading to:

$$[\theta]_\lambda^H = f_H[\theta]_\lambda^{H(\infty)}[1 - (k/n)] + f_\beta[\theta]_\lambda^\beta + f_R[\theta]_\lambda^R . \tag{77}$$

Here $[\theta]_\lambda^{H(\infty)}$ is the theoretical mean residual ellipticity of a 100% helix protein with infinite chain length. They proposed that $[\theta]_\lambda^{H(\infty)} = -39\,500$ at 222 nm and k

Table 5. The CD and ORD of helix, β sheet, and unordered forms based on five proteins [103]

[θ]^a				[m]^a			
λ (nm)	H	β	R	λ (nm)	H	β	R
190	70,100	−2,880	−20,300	197	64,100	−2,960	−5,130
193	77,000	12,900	−36,000	200	76,400	15,100	−21,600
196	68,200	14,000	−37,900	203	69,900	12,200	−20,900
199	37,200	6,050	−23,200	206	−56,500	7,600	−18,500
201	16,800	8,810	−23,300	209	37,300	4,520	−13,000
204	−9,120	300	−11,300	212	27,800	3,710	−11,800
207	−22,300	−4,320	−5,770	215	22,300	−230	−9,520
210	−26,400	−8,190	−2,200	218	15,800	−4,380	−6,220
213	−24,800	−8,680	−850	221	9,620	−8,560	−3,630
216	−26,600	−9,210	1,230	224	2,080	−9,580	−2,280
219	−28,900	−6,890	1,720	227	−5,320	−8,660	−1,320
222	−30,000	−3,360	1,580	230	−11,600	−5,020	−1,360
225	−28,700	1,540	260	233	−13,800	−1,900	−1,740
228	−24,000	4,390	−480	236	−13,500	−1,090	−1,010
231	−17,300	4,580	−770	239	−11,700	−940	−670
234	−11,300	3,540	160	242	−9,340	−690	−950
237	−6,250	2,410	−90	245	−7,300	−330	−1,230
240	−2,950	3,370	−1,040	248	−5,650	−350	−1,340
243	−1,230	2,040	−930	251	−4,390	−470	−1,390

^a Dimensions: (deg cm²) dmol⁻¹

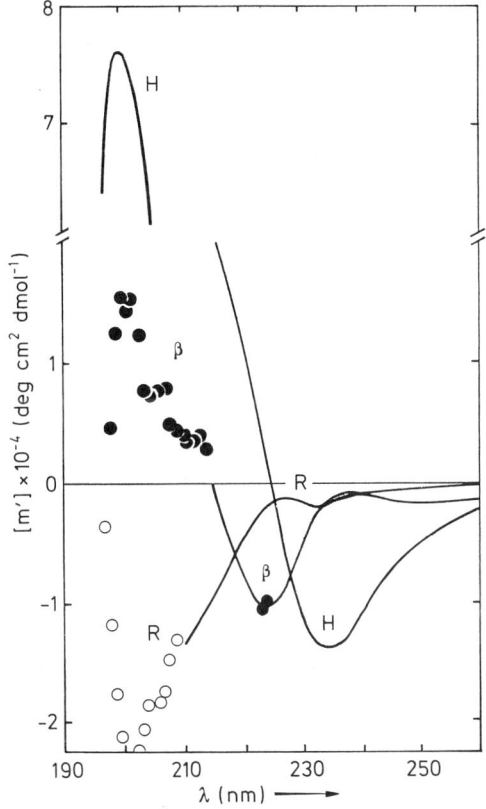

Fig. 33. CD of the helix (H), β sheet (β), and unordered (R) form computed from the CD of five proteins [103]

= 2.57 at 222 nm. Table 5 lists the computed $[\theta]$ and $[m']$ values of helix, β sheet, and random coil structure from the experimental data of five proteins (myoglobin, lysozyme, lactate dehydrogenase, papain, and ribonuclease). Figures 33 and 34 show the corresponding spectra of three conformations. In these figures points indicate deviations from the curve. The CD extrema of the helix are located at 222 (negative), 209 (negative), and 192 (positive) nm and the ORD extrema at 233 (negative) and 200 (positive) nm. These extrema can be a measure for the estimation of the helix content. The CD of the β sheet has a trough at 216 nm and two positive bands at 225 and 240 nm; this situation is quite delicate, however. Hatano and Yoneyama [102] estimated the helix contents of various kinds of synthetic poly(α amino acids) without side-chain chromophores from the observed values of $[\theta]_{222}$, Moffitt's b_0, and $[m']_{233}$. Figures 35–37 show that there is some discrepancy which may arise from the appreciable content of β structures in poly(L-lysine) with relatively low molecular weight and poly(L-α, γ-diaminobutyric acid) which preferably form a β sheet structure.

Pocker and Biswas [104] used a simpler way to estimate the helix content of insulin from the CD spectrum. An examination of the computed spectra of the α helix, β sheet, and random coil of Chen et al. [103] shows that the CD contribution of β sheet and random coil to the mean residue ellipticity at 223 nm (see Fig. 33) of a protein is negligibly small. The $[\theta]_{223}$ values for the monomer and dimer of insulin are —6.07

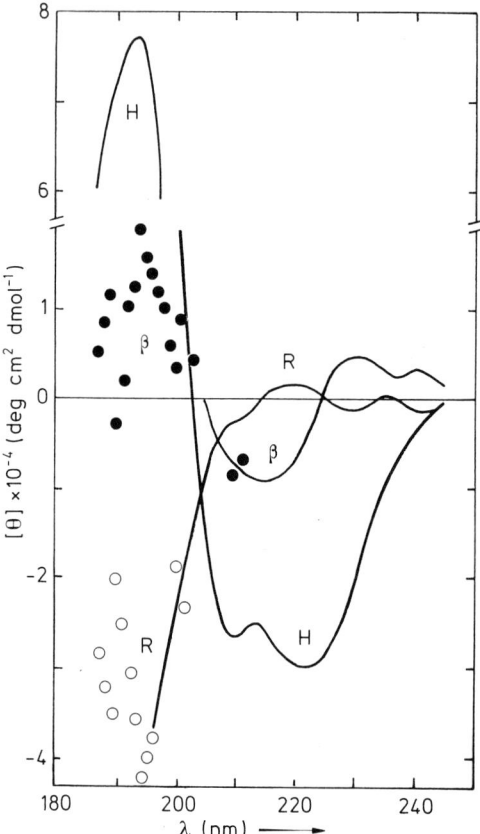

Fig. 34. ORD of the helix (H), β sheet (β), and unordered (R) form computed from the ORD of five proteins [103]

$\times 10^3$ and -11.30×10^3, respectively, which are corresponding to 24.2 and 45% helical conformation.

5.2 Prediction of Secondary Structure in Proteins

The CD spectrum of a protein is a direct reflection of its secondary structure. A method is presented for predicting the secondary structure of a protein from its CD spectrum. According to the Chou and Fasman method [105, 106], $\langle P_\alpha \rangle$, $\langle P_\beta \rangle$, and $\langle P_t \rangle$, that is, the average potentials for any protein segment to be in an α helix (α), β sheet, and β turn (t), respectively, are computed for predicting the secondary structure on the basis of the X-ray data of fifteen proteins. Table 6 shows these parameters for 20 amino acid residues. Recently, we predicted the secondary structure of *Aplysia* myoglobin, the oxygenated form of which is extremely unstable, from its amino acid sequence [107] using the Chou-Fasman method. Figure 38 shows that *Aplysia* myoglobin is composed of 144 amino acid residues and contains a single histidine residue at position 95 that most likely corresponds to the heme-binding site. The segments of β turns are assigned by computing the relative probability of β turns, P_t, since the

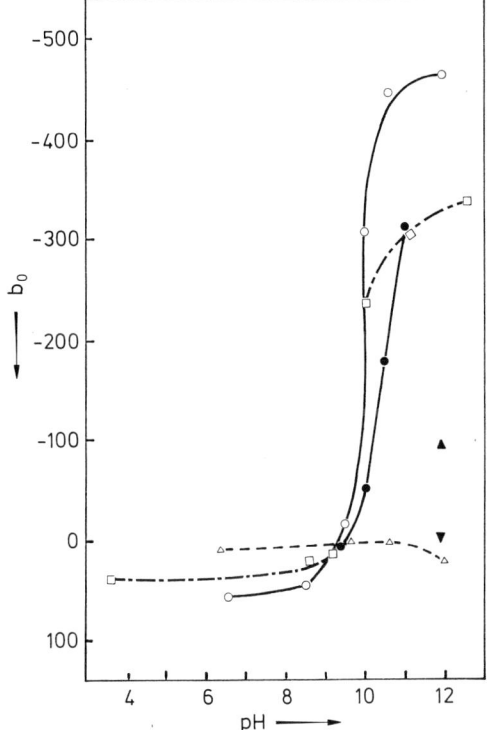

Fig. 35. The pH dependence of Moffitt's b_0 values of poly(α-amino acids) in aqueous solution [102]:
○: poly(L-lysine) (160),
●: poly(L-lysine) (46),
□: poly(L-ornithine) (190),
△: poly(L-α,γ-diaminobutyric acid) (110).
The numbers in parentheses are degrees of polymerization. The b_0 values in the case of addition of salts to poly(α,γ-diaminobutyric acid) solution at pH are indicated. ▲, 2.0 M KSCN; ▽, 2.0 M KCl

correct prediction of chain reversal will lead to a greater understanding of the tertiary folding. Table 7 indicates that there are six α helical segments, two β sheet domains, and four parts of β turns. This conformational analysis reveals that the fraction of each form is as follows: $f_\alpha = 0.44$, $f_\beta = 0.07$, f_t (β turn) = 0.11, and f_r (random coil) = 0.38. Figure 39 shows the CD spectrum of *Aplysia* myoglobin, and it is obvious that the CD magnitude at 222 nm is smaller ($-18\,500$ deg cm^2 dmol^{-1}) than that of sperm whale myoglobin ($-24\,000$ deg cm^2 dmol^{-1}). The extent of the contribution of the β sheet to the overall CD magnitude at 222 nm is very small and was calculated to be -123 deg cm^2 dmol^{-1}. CD analysis allows to estimate the α helical fraction f_α which is in the range of 0.45–0.50. This value is approximately equal 0.44, as calculated from the Chou-Fasman method, within our experimental errors. From the above-mentioned analysis, a predicted secondary structure of *Aplysia* myoglobin can be depicted as shown in Fig. 40. Each helical loop represents a single turn consisting of 3.6 amino acid residues, and the chain reversals at a β turn tetrapeptide unit. Both the β turn at positions 99–102 and the proline bends at positions 113, 122, and 123 are depicted to allow an antiparallel ordering between the β_x domain and the β_y domain. If the heme iron is coordinated by a single histidine residue at position 95, this predicted structure reveals that the heme moiety is open to a less hydrophobic region. This situation is quite different from that in mammalian myoglobin, the oxygenated form of which is stable. For example, the half-life of the oxygenated form is 5 hr for *Aplysia*, but 10 days for bovine, and 35 days for sperm whale myoglobin, at pH 9.2.

Table 6. Frequency of β-turn residues in 15 proteins, and comparison of β-turn conformational parameters P_t for 20 amino acids with their P_α and P_β values [105]

Amino Acid	n^a	n_t^b	f_t^c	P_t^d	P_α^e	P_β^e
Gly	222	99	0.446	1.68	0.53	0.81
Asn	139	62	0.446	1.68	0.73	0.65
Ser	201	83	0.413	1.56	0.79	0.72
Pro	81	33	0.407	1.54	0.59	0.62
Asp$^{(-)}$	102	34	0.333	1.26	0.98	0.80
Tyr	118	39	0.331	1.25	0.61	1.29
Cys	45	14	0.311	1.17	0.77	1.30
Trp	44	13	0.295	1.11	1.14	1.19
Lys$^{(+)}$	150	40	0.267	1.01	1.07	0.74
Arg$^{(+)}$	79	21	0.266	1.00	0.79	0.90
Thr	162	43	0.265	1.00	0.82	1.20
Phe	64	12	0.188	0.71	1.12	1.28
His$^{(+)}$	60	11	0.183	0.69	1.24	0.71
Met	28	5	0.179	0.67	1.20	1.67
Ile	118	18	0.153	0.58	1.00	1.60
Ala	204	31	0.152	0.57	1.45	0.97
Gln	101	15	0.149	0.56	1.17	1.23
Leu	156	22	0.141	0.53	1.34	1.22
Glu$^{(-)}$	94	11	0.117	0.44	1.53	0.26
Val	175	14	0.080	0.30	1.14	1.65
Total	$N = 2343$	$N_t = 620$	$\langle f_t \rangle^f = 0.265$	$\langle P_t \rangle^g = 1.00$	$\langle P_\alpha \rangle^g = 1.00$	$\langle P_\beta \rangle^g = 1.00$

[a] n = total occurrence of each residue in the 15 proteins;
[b] n_t = total occurrence of each residue in the β turns;
[c] $f_t = n_t/n$ is the frequency of residues in β turn regions;
[d] $P_t = f_t/\langle f_t \rangle$ is the conformational parameter for the β turn;
[e] See text. [f] $\langle f_t \rangle = N_t/N$ is the average frequency of all residues in the β-turn regions. [g] See text

The finding suggests that *Aplysia* myoglobin may have a heme environment to allow more easier attack by H_2O molecules or hydroxide anions upon FeO_2 bonding, resulting in a very rapid formation of metmyoglobin. In mammalian myoglobin the hydrophobic heme environment and the distal histine residue may stabilize the oxygenated form, whereas the less hydrophobic heme pocket and the lack of the distal histidine in *Aplysia* myoglobin may result in lowering of its oxygenated form.

As shown in Figs. 33 and 39, the structural contributions to the CD spectrum in the range of 200–250 nm is quite important for predicting the secondary structure of a protein. However, β sheet and β turn structures are strongly reflected in the CD spectrum magnitude in the vacuum ultraviolet region. Hennessy and Johnson, Jr. further extended the observation range of CD spectra into the vacuum ultraviolet to 165 nm [108]. They found that the positive band at 190 nm of a helical conformation, shifts from 185 nm (in elastase) to 195 nm (in cytochrome *c*) with a 50-fold fluctuation in magnitude. They used an eigenvector method of multicomponent matrix analysis with a set of 15 protein CD spectra and one polypeptide CD spectrum over the range of 178–260 nm to generate orthogonal CD spectra as a basis. The β sheet structures were divided into two classes, parallel and antiparallel. The contribution of β turns was

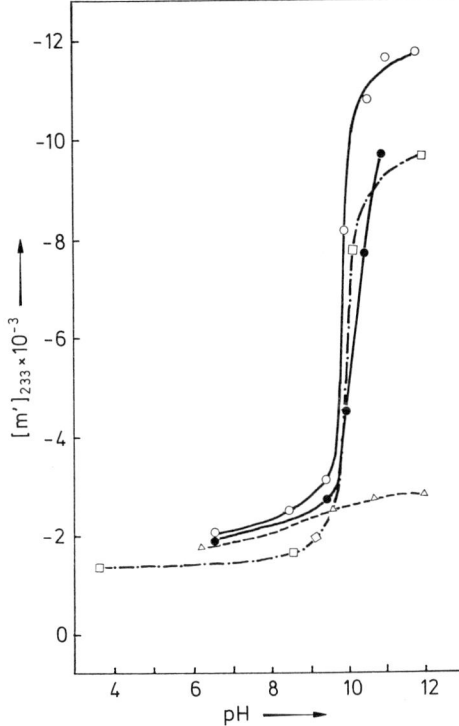

Fig. 36. The pH dependence of $[m']_{233}$ of poly(α-amino acids) in aqueous solution [102]. Refer to Fig. 35 for the meaning of the numbers in parentheses.
○: poly(L-lysine) (160),
●: poly(L-lysine) (46),
□: poly(L-ornithine) (190),
△: poly(L-α,γ-diaminobutyric acid) (110)

classified into four types on the basis of the classification criteria of Chou-Fasman [109]. This method provides a more precise prediction of the secondary structure of a protein. However, a protein with a CD contribution from aromatic chromophores will be not well analyzed.

Manavalan and Johnson, Jr. [96] presented typical CD spectra of proteins divided into four classes on the basis of their secondary structures: all-α-helical (predominantly α-helical), all-β-sheet (predominantly β-sheet), α + β (α-helix and β-sheet domains exist separately), and α/β (intermixed modules of α helix and β sheet). This diagnosis is very convenient to predict the secondary structure of a protein. All-α proteins exhibit a characteristic α-helix CD spectrum (exemplified by that of myoglobin; Fig. 41a having two distinct negative troughs at 208 and 222 nm, a positive peak in the region of 190–195 nm, and a cross-over point near 172 nm. Myoglobin, hemoglobin, parvalbumin, cytochrome b_{562}, cytochrome c, hemerythrin, and polyglutamic acid belong to this class. The α + β class of proteins that are dominated by the α-helix portion exhibits two negative troughs at 208 and 222 nm and a positive peak in the region of 190–195 nm. The negative CD magnitude at 208 nm is often larger than that of the band at 222 nm, and this difference among them is reverse to that in cases of α-helical proteins. Furthermore, the cross-over point from positive to negative CDs is above 172 nm, as shown in Fig. 41b. This feature differentiates the α + β proteins from the α-helical proteins. Lysozyme, ribonuclease A, papain, insulin, cytochrome c peroxidase, thermolysin, and *Staphylococcus* nuclease are grouped with the α + β class. The α/β group of proteins has more pronounced 222 nm bands than the α + β class, as shown in Fig. 41c. In some cases, larger overlapping of the 208 nm band over the 222 nm band reduces to a single broad trough. In the α/β group, the cross-over point varies in the region of 175–185 nm. Subtilisin BPN', lactate dehydrogenase, triose phosphate isomerase, subtilisin novo, glyceraldehyde-3-phosphate dehydrogenase, flavodoxin, pyruvate kinase, aldolase, hexokinase, glutathione reductase, thioredoxin, phosphoglycerate kinase, and dihydrofolate reductase exhibit CD spectra

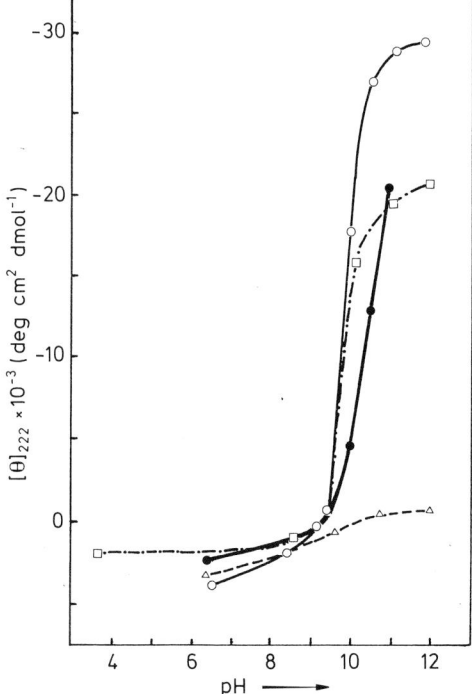

Fig. 37. The pH dependence of $[\theta]_{222}$ of poly(α-amino acids) in aqueous solution [102].
Refer to Fig. 35 for the meaning of the numbers in parentheses.
○: poly(L-lysine) (160),
●: poly(L-lysine) (46),
□: poly(L-ornithine) (190),
△: poly(L-α,γ-diaminobutyric acid) (110)

```
              1                 5                    10                    15                  20
Acetyl-Ser-Leu-Ser-Ala-Ala-Glu-Ala-Asp-Leu-Val-Gly-Lys-Ser-Trp-Ala-Pro-Val-Tyr-Ala-Asn-
                              25                    30                    35                  40
        Lys-Asp-Ala-Asp-Gly-Ala-Asn-Phe-Leu-Leu-Ser-Leu-Phe-Glu-Lys-Phe-Pro-Asn-Asn-Ala-
                              45                    50                    55                  60
        Asn-Tyr-Phe-Ala-Asp-Phe-Lys-Gly-Lys-Ser-Ile-Ala-Asp-Ile-Lys-Ala-Ser-Pro-Lys-Leu-
                              65                    70                    75                  80
        Arg-Asp-Val-Ser-Ser-Arg-Ile-Phe-Thr-Arg-Leu-Asn-Glu-Phe-Val-Asn-Asn-Ala-Ala-Asp-
                              85                    90                    95                 100
        Ala-Gly-Lys-Met-Ser-Ala-Met-Leu-Ser-Gln-Phe-Ala-Ser-Glu-His-Val-Gly-Phe-Gly-Val-
                             105                   110                   115                 120
        Gly-Ser-Ala-Gln-Phe-Glu-Asn-Val-Arg-Ser-Met-Phe-Pro-Ala-Phe-Val-Ala-Ser-Leu-Ser-
                             125                   130                   135                 140
        Ala-Pro-Pro-Ala-Asp-Asp-Ala-Trp-Asn-Lys-Leu-Phe-Gly-Leu-Ile-Val-Ala-Ala-Leu-Lys-

        Ala-Ala-Gly-Lys
```

Fig. 38. Amino acid sequence of myoglobin from *Aplysia kurodai* [107]

Table 7. Conformational parameters for all residues of *Aplysia* myoglobin [107]

Residues	$\langle P_\alpha \rangle$	$\langle P_\beta \rangle$	$\langle P_t \rangle$		Predicted form with notation
1	0.79	0.72			c
2– 10	1.27	0.97			α_A
11– 14	0.88	0.86	1.34	$1.6 \cdot 10^{-4}$	β turn
15– 27	1.01	0.91			c
28– 36	1.19	1.02			α_B
37–38	0.66	0.63			c
39– 42	0.88	0.89	1.29	$1.1 \cdot 10^{-4}$	β turn
43– 46 [a]	1.16	1.08	0.81	$0.6 \cdot 10^{-5}$	c
47– 50	0.86	0.75	1.31	$0.3 \cdot 10^{-4}$	β turn
51– 66	1.01	0.97			c
67– 71	1.01	1.24			β_X
72	0.73	0.65			c
73– 96	1.15	0.93			α_C
97– 98	0.82	1.04			c
99–102	0.74	0.99	1.30	$0.5 \cdot 10^{-4}$	β turn
103–108	1.19	1.00			α_D
109–113	0.89	1.03			c
114–121	1.19	1.06			α_E
122–130	0.99	0.81			c
131–136	1.07	1.29			β_Y
137–144	1.22	0.92			α_F

[a] Residues 43–46 are not predicted

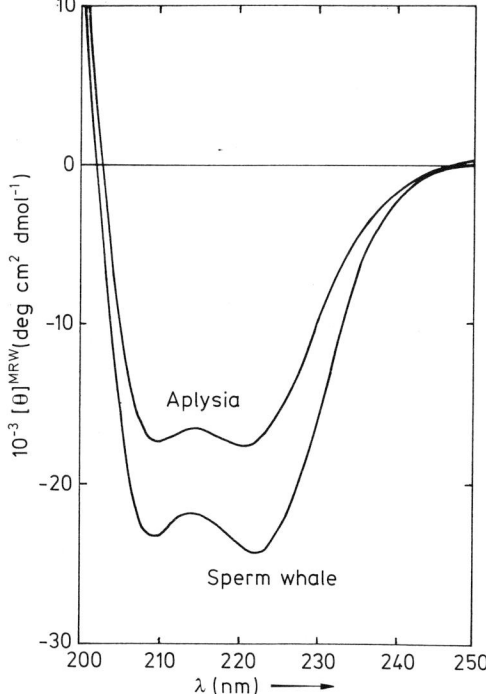

Fig. 39. CD spectrum of *Aplysia* myoglobin compared to that of sperm whale myoglobin in 0.01 M phosphate buffer, pH 7.4 [107]

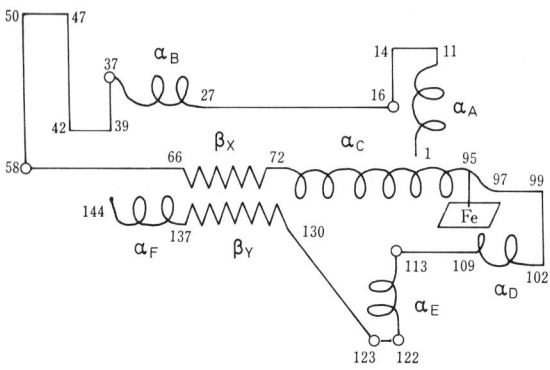

Fig. 40. A predicted secondary structure of *Aplysia* myoglobin [107].
ꝏꝏ : α-helical structure.
⋀⋀ : β-sheet structure.
── : random-coil structure.
▯▯ : β turn.
Conformation boundary residues are numbered as well as the six proline residues represented by ○

grouped with the α/β class. All-β proteins exhibit CD spectra characterized by the absence of peaks found for α-helical proteins. In this class of proteins, the CD spectrum in this region is low in magnitude, and demonstrates a variety of shapes when the contribution of β turns and others to the CD is appreciable. α Cobaratoxin, Bence-Jones proteins, gene 5 protein, immunoglobulin, prealbumin, superoxide dismutase, and concanavalin A exhibit CD spectra with a trough in the range of 210–220 nm (see Fig. 41 d). This class of proteins has a negative trough in the region of 170–180 nm. On the other hand, soybean trypsin inhibitor, wheat germ agglutinin, rubredoxin, elastase, and α-chymotrypsin have a CD trough near 200 nm and a positive peak in the range of 180–190 nm, as shown in Fig. 41 e. This characteristic results from the occurrence of very short β sheets. Provencher and Glöckner [110] also emphasized the need to observe the CD spectrum in the vacuum ultraviolet region to predict the secondary structure of a protein having β sheet(s).

For the past few years, complete amino acid sequences have been determined from the nucleotide sequences of structural genes.

The knowledge of amino acid sequences of proteins may provide their three-dimensional structures if CD predictions are valid. However, the CD prediction for the secondary structure of a given protein, especially having β-sheet structures, is not valid enough to probe the structure-function relationships in native proteins. We are awaiting the development of a more precise prediction method for secondary structures of proteins.

Membrane-bound proteins extend from the cytoplasmic membrane. Analysis by the method of Kyte and Doolittle [186] is quite useful for predicting the protein segments extending into the cytoplasmic membrane. The Kyte and Doolittle method, the so-called *hydropathic index* method, if it is coupled with the Chou-Fasman method, safely differentiates the protein segment which is located outside the membrane, from the helices within the membrane. The best examples are cytochrome P-450 [187], cytochrome b_5 [188], reaction centers [189,190] and light-harvesting protein complexes [191] of photosynthetic bacteria. This hydropathy index also leads to estimate the assumed relative orientation of functional chromophores, such as bacteriochlorophylls or heme moieties, if the observed CD, based on the relative orientation of the chromophores, is calculated in terms of molecular orbital approximation [192].

5.3 Conformational Change of Proteins

In general proteins and synthetic poly-α-amino acids can form ordered structures through the inter- and intra-hydrogen bonds among the amide groups in their back-

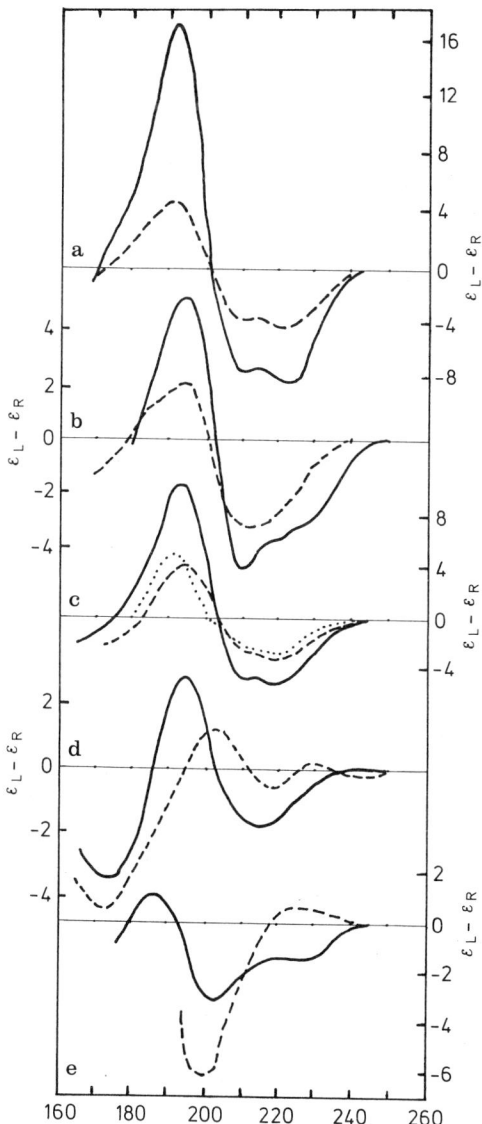

Fig. 41 a–e. Typical CD spectra of proteins [96].
a. All-α proteins; myoglobin (———); cytochrome c (– – –).
b. α + β proteins; lysozyme (———); ribonuclease A (– – –).
c. α/β proteins; triose phosphate isomerase (———); flavodoxin (– – –); subtilisin BPN' (......).
d. all-β proteins; prealbumin (———); Bence-Jones protein (– – –).
e. all-β proteins; α-chymotrypsin (———); soybean trypsin inhibitor (– – –).
Reprinted by permission from Nature (London), Vol. 305, pp. 831 (1983), Copyright ©, Macmillan Journals Limited)

bone structures. When the side-chain groups are ionized in proteins and poly(α-amino acids), the static charges on the side chains reinforce to convert to their random-coiled structure. Figures 35–37 exemplify the conformational change from α-helical to random-coil structure upon raising the pH in solution. This conformational change reflects the pK value of the ionizable group in the side chain.

Quantitative analyses of the conformational change in proteins have been carried out by many authors [111]. Above all, the unfolding, conformational change from α-helix to random coil of α-lactalbumin has been studied in detail [112–114]. The transition between native and acid-denatured states was followed kinetically by the stopped-flow pH-jump method showing a rapid change in absorption [113] and CD [114]. The native human α-lactalbumin exhibits three CD bands in the near ultraviolet region, at 270, 293, and 299 nm (Fig. 42); these bands diminish in magnitude at acidic pH, though the CD magnitude at 222 nm remains about two-thirds of that of the native one. Therefore, the aromatic side chains buried in the interior of the molecule have an increased rotational freedom upon raising the pH, although the helical conformation of the main peptide chain remains undestroyed. The thermal unfolding process of α-lactalbumin also has been studied by kinetic measurements of the Joule heating temperature-jump.

More dramatical examples of conformational change have been reported for Ca^{2+}-binding proteins. Osteocalcin is an abundant Ca^{2+}-binding protein characterized by the presence of the unique vitamin K-dependent γ-carboxyglutamic acid [115]. The

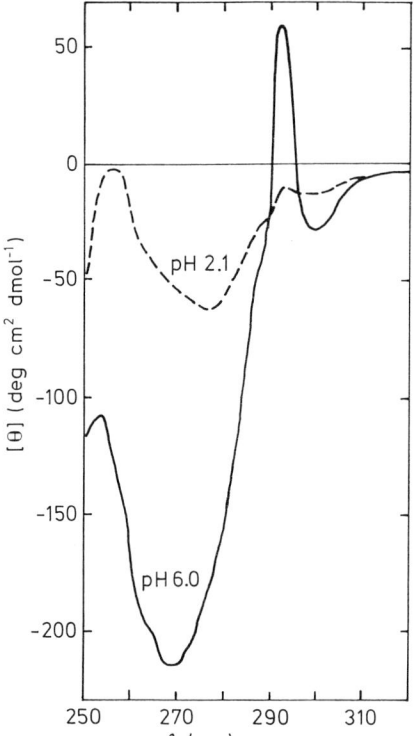

Fig. 42. CD spectral change of human α-lactalbumin [114]. Solid line: pH 6.0. Broken line: pH 2.1

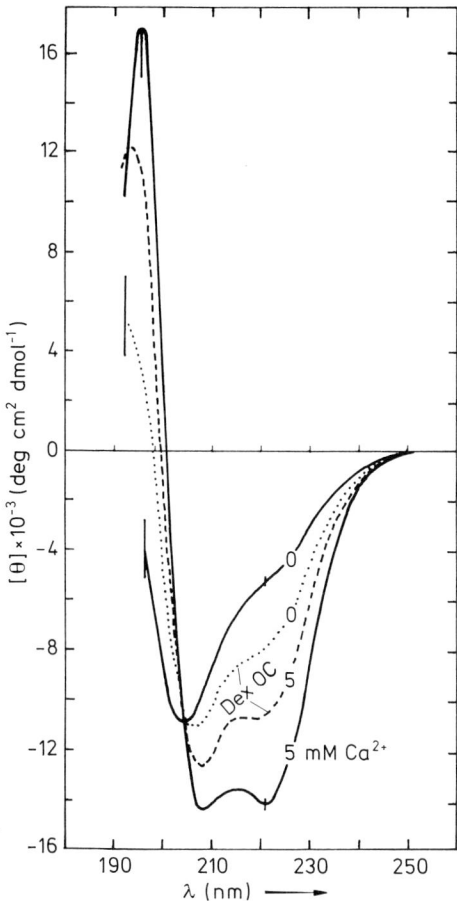

Fig. 43. CD spectra of native and decarboxylated chicken osteocalcin in the presence and absence of 5 mM Ca^{2+} [115].
———, 0: apoosteocalcin.
———, 5: osteocalcin with 5 mM Ca^{2+}.
......., 0: decarboxylated apoosteocalcin.
– – –, 5: decarboxylated osteocalcin with 5 mM Ca^{2+}

malonic acid moiety of this amino acid participates directly in metalcation binding. CD spectra reveal that a drastic conformational change occurs when Ca^{2+} ions are bound to the protein (see Fig. 43). Ca^{2+}-free osteocalcin is a random coil with only 8% of its residues in the α helix. In the presence of 5 mM Ca^{2+}, 38% of the protein residues adopt the α-helical conformation. The transition midpoint for the α helix occurs at 0.75 mM Ca^{2+}. The protein exhibits a weak CD spectrum in the near ultraviolet region without Ca^{2+}. When Ca^{2+} is added to the protein, the negative CD decreases dramatically. This suggests that tyrosine and phenylalanine residues are fixed to keep a rigid geometry around the main chain of the peptide. The appearance of the positive CD at 295 nm indicates that Ca^{2+} may induce phenolate anion formation (Fig. 44).

A similar conformational change has been detected for calmodulin; a heat-stable Ca^{2+}-binding protein. The field of Ca^{2+} research continues to expand noticeably, with articles focusing on calmodulin [116]. Calmodulin regulates the enzymatic activities of various enzymes such as adenylate cyclase and cyclic nucleotide phosphodiesterase. There are four Ca^{2+}-binding sites in calmodulin, and the Ca^{2+} binding

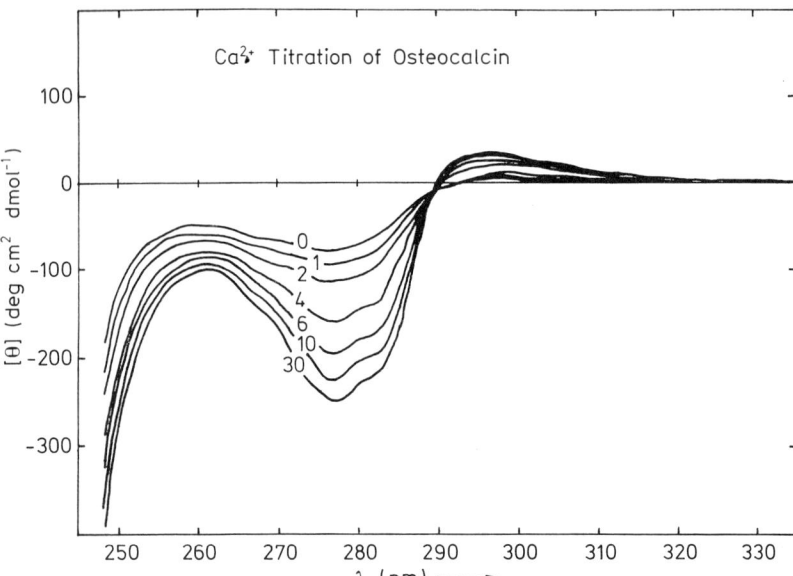

Fig. 44. Near ultraviolet CD spectral change of osteocalcin at increasing concentrations of Ca^{2+} [115]. Apoosteocalcin (1.6 mg/ml in 150 mM NaCl — 20 mM Tris-HCl, pH 7.4) was titrated. The numbers indicate the concentration of Ca^{2+} in mM

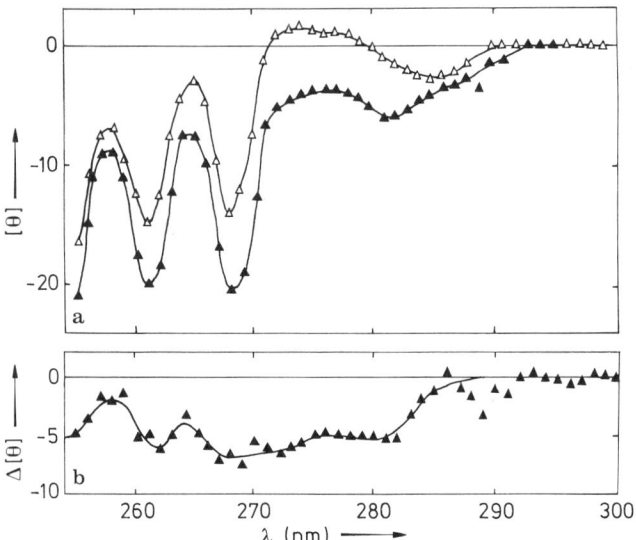

Fig. 45a and b. Effect of KCl (0.1 M) on the near ultraviolet CD of calmodulin [117].
a. CD of calmodulin (3.5–5 mg/ml) in HEPES (0.01 M) — KOH buffer, pH 7.5, containing 10^{-6} M EDTA.
—△—: without KCl.
—▲—: after addition of KCl (0.1 M).
b. The difference in ellipticity ($\pm$KCl)

induces the protein to form a type of ordered structure, which is responsible for activating the different enzyme systems. The conformational change from random coil to an ordered structure was investigated in detail in the presence or absence of Ca^{2+} ions [117]. Experiments were performed with purified bovine brain calmodulin. With 2–5 mg of protein/ml, the ratio of Ca^{2+} to calmodulin concentration was adjusted to 4 mol Ca^{2+}/mol of protein. A large CD change between 250 and 290 nm was found, as shown in Fig. 45, when 0.1 M KCl was added. Addition of 0.1 M KCl in the absence of Ca^{2+} resulted in negative CD extrema in the region of 250–280 nm. Binding of Ca^{2+} to calmodulin in the presence of 0.1 M KCl induced an additional change of negative ellipticity in the region of 250–290 nm. This change is shown in Fig. 46. The increase indicates that Ca^{2+} affects the environment of both tyrosyl (279 and 286 nm) and phenylalanyl moieties (261 and 268 nm). Changes in ellipticity at 279 and 268 nm and absorption at 279 nm indicate that the conformational changes are almost completed upon binding of 2 mol of Ca^{2+}/mol of calmodulin (Fig. 47). Sigmoidicity of the Ca^{2+}-dependent spectral changes is parallel to that of the stimulation of phosphodiesterase activity. The CD change is due to the increase of the Ca^{2+} concentration in the first stage of Ca^{2+} binding. However, the binding of Ca^{2+} to the third

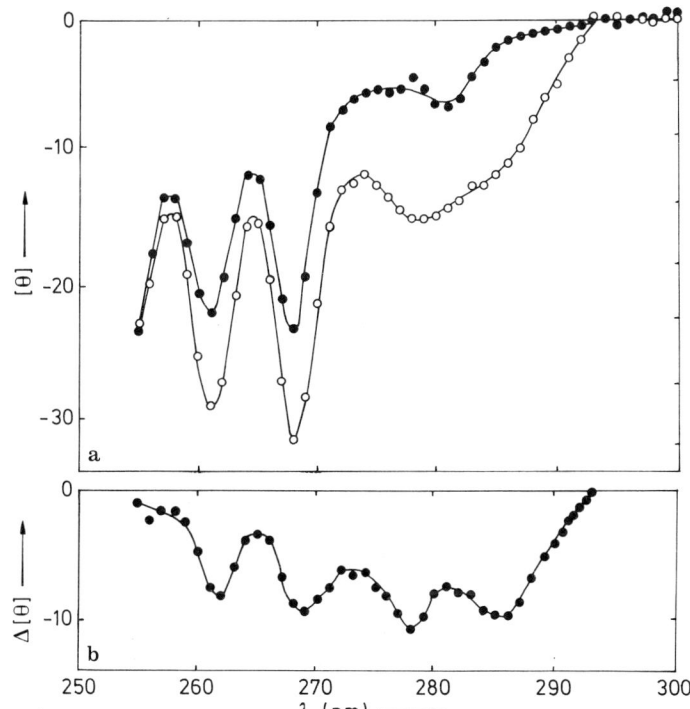

Fig. 46a and b. Effect of Ca^{2+} on the near ultraviolet CD of calmodulin [117].
a. CD of calmodulin (3.5–5 mg/ml) in HEPES (0.01 M) — KOH buffer, pH 7.5, containing 0.1 M KCl and 10^{-6} M EDTA.
—●—: without Ca^{2+}.
—○—: after addition of $CaCl_2$ of 1.3×10^{-3} M.
b. The difference in ellipticity ($\pm CaCl_2$)

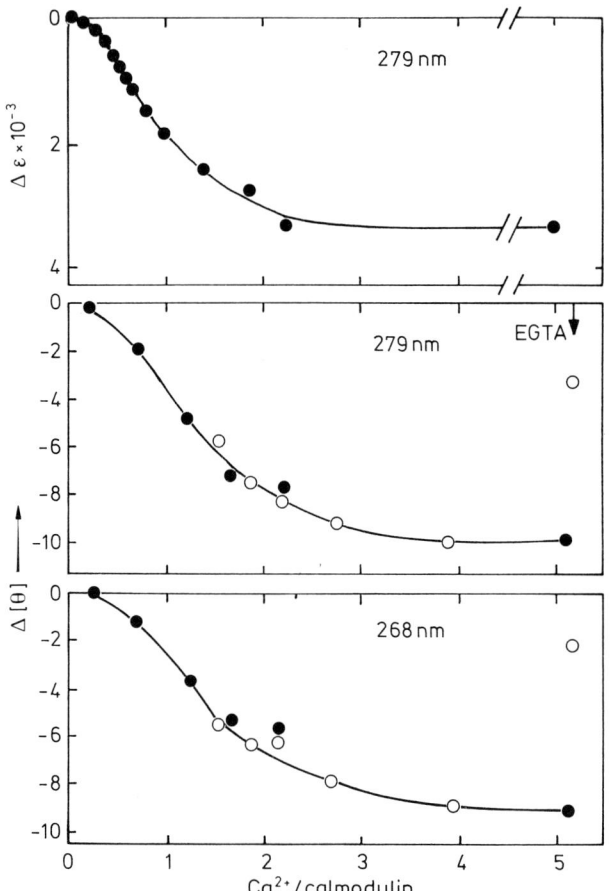

Fig. 47. Changes in the extinction coefficient at 279 nm and ellipticity at 279 and 268 nm as a function of bound Ca^{2+} [117].
The open and closed circles represent two different experiments

and fourth sites results in less additional changes in the environment of the tyrosyl and phenylalanyl moiety. Further changes of the conformation, from the Ca^{2+} binding to the third and fourth sites, is detectable by induced CD [118] and ^{19}F NMR [119] for an antagonist (trifluoperazine) to calmodulin and ^{43}Ca NMR for Ca^{2+} ions bound to the protein [120].

5.4 Induced Circular Dichroism in Side-Chain Chromophores

In the preceding Section (5.3) the CD bands in the near ultraviolet region are assigned to those of tyrosyl and phenylalanyl residues, whose thermal motions are restricted sterically keeping their geometry constant around the main chain in the proteins. The CD spectrum of aromatic moieties provides a good measure for the analysis of motional freedom around a given amino acid residue. Better examples have been reported on several kinds of amino acids and their polymers.

Optically active phenylglycine exhibits a larger CD than phenylalanine [121]. The absolute value of the optical anisotory factor $|\Delta\varepsilon/\varepsilon|$ of D-phenylglycine near 250 nm ($^1B_{2u} \leftarrow {}^1A_{1g}$ transition for benzene) is about eight times as large as that of L-phenylalanine at room temperature. The dissymmetric perturbation of the phenyl chromophore in D-phenylglycine can be regarded as more effective than in L-phenylalanine, since we can assume that the $|\Delta\varepsilon/\varepsilon|$ value is directly proportional to the effective strength of the perturbation by the dissymmetric field caused by the optically active moiety. This probably reflects the fact that the distance between the phenyl chromophore and the chiral $^\pi C$ carbon atom in D-phenylglycine is shorter than that in L-phenylalanine and the fact that the phenyl chromophore in D-phenylglycine rotates less freely than in L-phenylalanine. The rotational freedom of the phenyl chromophores both of D-phenylglycine and L-phenylalanine will be reduced at low temperature, while the distance between the phenyl chromophore and the chiral $^\pi C$ carbon atom of these compounds are not altered by the temperature decrease. The $|\Delta\varepsilon/\varepsilon|$ values for D-phenylglycine and L-phenylalanine are observed to be much enhanced at the temperature of liquid nitrogen compared to those at room temperature. From the findings we can say that the effective asymmetric perturbations by the chiral $^\pi C$ carbon atom are strengthened at lower temperatures as a result of the reduction of the rotational freedom of the phenyl group. The ratio of the $|\Delta\varepsilon/\varepsilon|$ value of D-phenylglycine to that of L-phenylalanine decreases from 8 to 2 by the temperature decrease from room temperature to that of liquid nitrogen. From this fact it is concluded that the difference between the $|\Delta\varepsilon/\varepsilon|$ values of D-phenylglycine and L-phenylalanine at room temperature is mainly attributed to the difference in the rotational freedom of the phenyl chromophore between D-phenylglycine and L-phenylalanine.

A similar CD has been observed for poly(γ-benzyl-D-glutamate) in concentrated solution forming a liquid crystalline phase [122]. In the concentration range of poly(γ-benzyl-D-glutamate) (PBDG) of less than 0.1 mol glutamyl residue/dm^3 in dichloromethane, no evidence for the occurrence of the higher order structure can be detected. ORD curves for such a concentration range show monotonic patterns as is encountered for the usual polypeptide solution having an α-helical structure. When the concentration is raised over 0.4 mol glutamyl residue/dm^3, microscopic examination reveals that a cholesteric liquid crystalline phase is formed. For this system, the ORD curve, giving an anomalous dispersion near 255 nm, is quite different from those of less concentrated solutions, and a CD band also was observed near 255 nm. The sign of the CD band is inverted from positive to negative when the solvent is replaced by tetrachloroethane. An extremely condensed phase is realized locally around closed-packed α helices lying in parallel equi-spaced planes making up the cholesteric helices, and the benzyl chromophores are imposed to take a definite arrangement relative to the plane. This means that the side-chain benzyl chromophores assume to keep a definite orientation relative to the cholesteric helix axis. The solvent-dependent inversion of the CD sign may be explained by the solvent dependency of the sense of the cholesteric helix reported by Robinson [123].

On the other hand, the dilute solution of poly(β-benzyl-L-aspartate) in chloroform or dichloromethane gives a weak but existant CD band centered at 252 nm. A large positive CD band at 222 nm means that poly(β-benzyl-L-aspartate) (PBLA) exists in the left-handed α-helical form. Poly(γ-p-nitrobenzyl-L-glutamate) also shows a definite negative CD band around 275 nm, even if the concentration of the polymer is quite low [122]. A more crowded arrangement of the benzyl group in PBLA around the main-chain α helix than in PBDG leads to an increase in the interaction between benzyl chromophores of PBLA, giving rise to the CD detected in PBLA, while such is not the case for PBDG. The introduction of the nitro group into the side-chain benzyl group of the polyglutamate intensifies the interaction between side-chain groups leading to an increase of probability of a definite orientation relative to the polypeptide main chain, because of the higher polarity of the p-nitrobenzyl chromophore. Again, as in the dilute solution of PBLA, there exists no Davydov pair of CD bands in concentrated solutions of PBDG, and so the benzyl chromophore of PBDG in the cholesteric liquid crystalline phase as well as that of PBLA in its dilute solution is not so closely stacked as to cause an interaction between electric transition moments in benzyl chromophores. A similar situation to that in PBLA occurs for various aromatic esters of poly(L-glutamic acid) [124–128] and poly(L-aspartic acid) [129–131]. For example, poly{γ-[2-(9-carbazolyl)ethyl]-L-glutamate} is of much interest to understand this situation [128]. As seen in Fig. 48, the intensity of the absorbance of poly{γ-[2-(9-carbazolyl)ethyl]-L-glutamate} (PCLG) is smaller than that of the corresponding absorbance of N-ethylcarbazole (EtCz). This can be interpreted to be caused by the remarkably small distance between carbazolyl chromophores in the

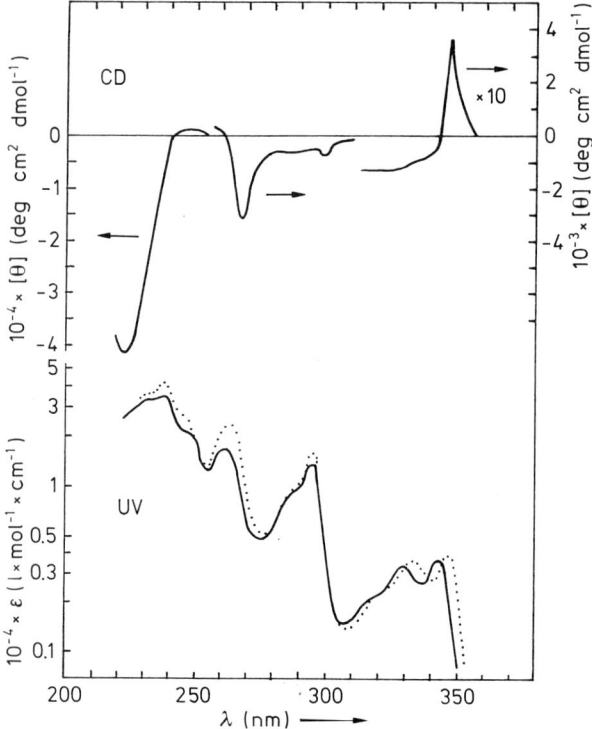

Fig. 48. Absorption and CD spectra of PCLG (———) and N-ethylcarbazole (·····) in 1,2-dichloroethane [128]

side chain of PCLG. However, the absorption of PCLG is larger than that of poly(vinyl carbazole) = (PNVC) [132], and no molecular-weight dependence on the absorption can be detected for PCLG, while PNVC shows a strong dependence of its molecular weight on the absorption. The difference in the magnitude or the molecular-weight dependence of the absorption between PCLG and PNVC can be ascribed to the larger distance between the carbazolyl chromophore and the main chain in PCLG than in PNVC. A Uniform correlation between the direction of the transition moments and the sign of the CD bands for the aromatic esters of poly(glutamic acid) and poly(aspartic acid), including PCLG, suggests that ICD bands, due to side-chain chromophores, originate from the pertubation by the α helix of the main chain to the side-chain chromophores, and that the side chain retains a substantially identical conformation in all of these polypeptides. Thus, the long-axis polarized band has the same sign as that of the CD band at 222 nm. The CD bands at 267 and 297 nm for PCLG are assigned to the long-axis polarized transitions of the carbazolyl group. The positive peak at 347 nm can be assigned to the short-axis polarized transition. This assignment has been confirmed by MCD measurement and molecular orbital calculation [133].

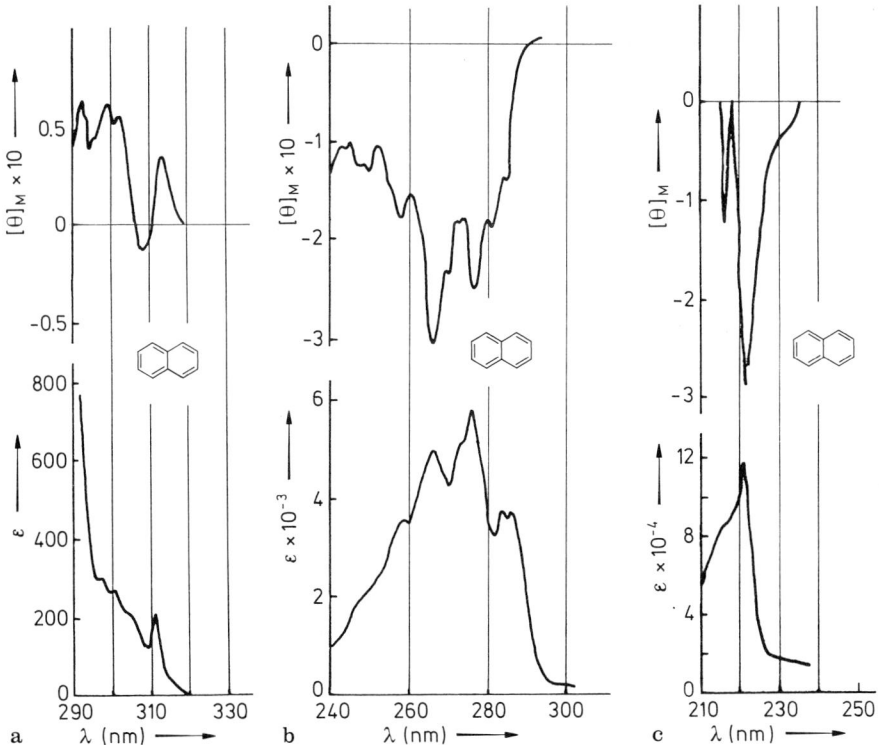

Fig. 49 a–c. MCD and absorption spectra of naphthalene [147].
a. ^1L_b band region.
b. ^1L_a band region.
c. ^1B_b and ^1B_a band regions

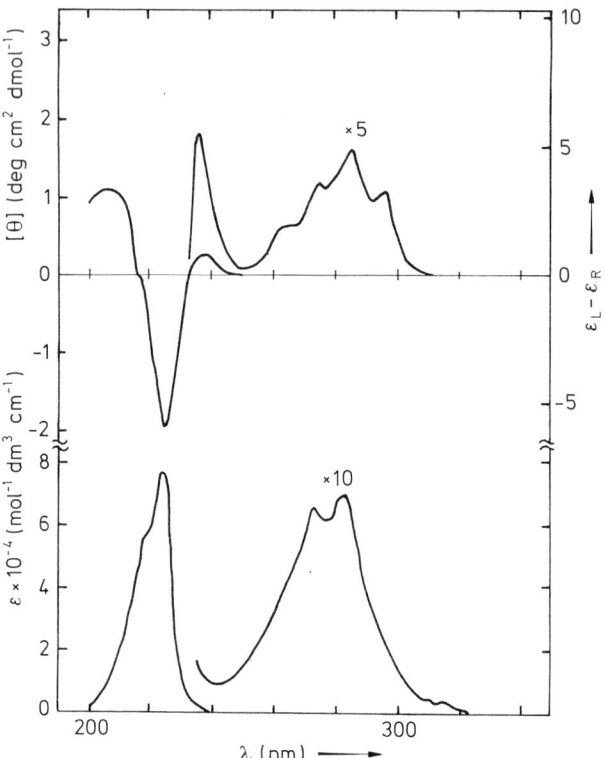

Fig. 50. Absorption and CD spectra of N-acetyl-L-1-naphthylalanine ethyl ester in trimethyl phosphate [146]

Contrary to the CD spectra of polyglutamates and polyaspartates, the CD spectra of poly(L-phenylalanine)[134–136], poly(L-tyrosine) [137,138], poly(L-tryptophan) [140,141], and the polypeptides of *p*-substituted phenylalanines [142–145] exhibit a Davydov-type splitting pair of the CD bands with opposite signs, indicating that the aromatic chromophores are regularly oriented relative to the α helix. The Davydov-type splitting in the CD bands results from the strong coupling of a given transition moment in the aromatic chromophores which are situated close to each other. The mechanism is the same as the one described in Sect. 3.2. Sisido, Egusa, and Imanishi reported a strong splitting of the CD bands observed in the ^1B_b ($B_{3u} \leftarrow A_g$, a long-axis polarized) transition of the naphthyl moiety in poly(L-naphthylalanine) [146]. According to MCD, naphthalene has B_{3u}, B_{2u}, B_{3u}, and B_{2u} transitions at 33 300, 36 200, 45 000, and 46 300 cm^{-1}, respectively [147]. Among them, $B_{3u} \leftarrow A_g$ and $B_{2u} \leftarrow A_g$ are long- and short-axis polarized transitions, respectively. The former is denoted as ^1B_b, the latter as ^1B_a. The MCD peak is sharp for the ^1B_a band, which is observed only as a shoulder in the usual absorption spectrum (Fig. 49). Figures 50 and 51 show absorption and CD spectra of N-acetyl-L-1-naphthylalanine ethyl ester and poly-(L-naphthylalanine) (PLNA) in trimethyl phosphate solution, respectively. The ^1B_b band exhibits a marked hypochromicity for PLNA as compared with N-acetyl-L-1-naphthylalanine ethylester. The CD spectrum of PLNA exhibits a marked exciton

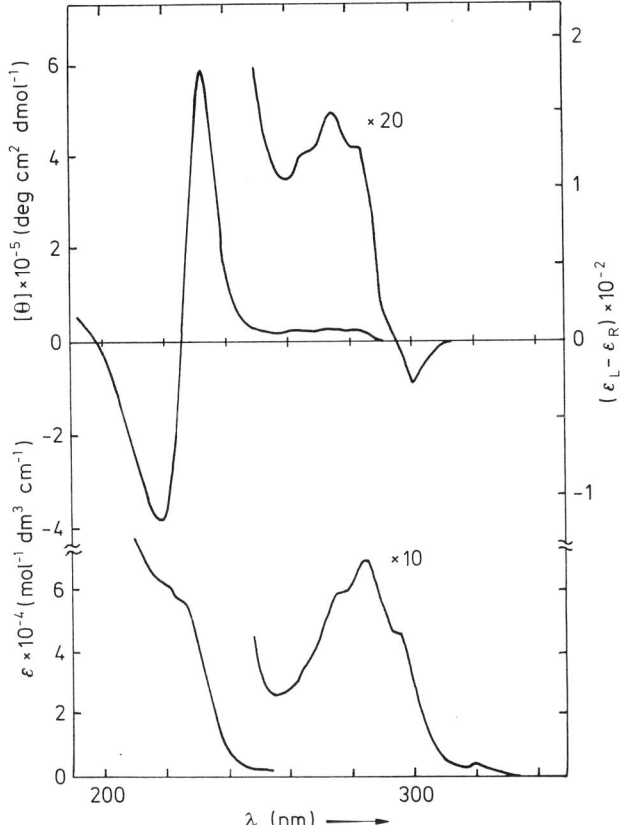

Fig. 51. Absorption and CD spectra of poly(L-naphthylalanine) in trimethyl phosphate [146]

splitting in the ^1B_b-band region. The large exciton coupling stands in contrast to the observation that poly(naphthylmethyl-L-glutamate) shows smaller CD peaks in the ^1B_b region by an order of magnitude without any significant splitting. The calculated CD curve proposes that PLNA in solution either has the conformation of a left-handed 3_{10} helix or right-handed δ helix. The former is energetically more stable, but the latter gives a more reasonable CD curve. On the basis of molecular display analysis, the face-to-face longitudinal distance between a pair of naphthyl groups, located in the vicinity of each other, is estimated to be about 3 Å for the former conformation and 4 Å for the latter. The distance is short enough to generate exciton coupling between the transition moments of the adjacent naphthalene chromophores. For D-homo-5α-androstane-3β,15β-bis(p-dimethylaminobenzoate), the distance between the two benzoate chromophores is estimated to be 12.8 Å from the calculation of the exciton coupling observed in its CD curve. In general, the CD amplitude is proportional to the reverse of the square of the distance between the chromophores generating the exciton coupling CD. The fixing of the mutual geometry between a pair of given chromophores and the magnitudes of their electric transition moments are of much importance for generating the exciton coupling CD.

5.5 Induced Circular Dichroism of Aromatic Compounds Bound to Proteins

Since Blout and Stryer [148] observed induced optical activity in the absorption region of acridine orange (AO) bound to the helical poly(α,L-glutamic acid (PLGA), many authors have discussed this induced optical activity [12]. The helical conformation is assumed to exist in the uncharged system of PLGA with AO, because the charged side chains in PLGA are neutralized with the ionic coupling between the carboxylate anion and AO cation. In the case of R/D (glutamyl residue-to-dye ratio) around unity, the polyanion of PLGA may be neutralized with AO cations, thus the PLGA-AO complex may take a helical form in a neutral and alkaline aqueous solution [149]. Figure 52 shows the absorption spectra of the PLGA-AO complex at pH 4.5, indicating that the α band at 492 nm (due to monomeric AO) becomes stronger than either the β band at 470 nm (due to dimeric AO) or the γ band at 450 nm (due to aggregated AO) with a decreasing R/D ratio. This fact indicates that only some of the carboxylic groups can bind to the dye. Excess dye remains free in solution, because of the low degree of ionization of the carboxylic groups. Figure 53 shows that molar ellipticities of the ICD bands at 435 and 465 nm decrease with decreasing R/D. Since the molar ellipticities were calculated on the basis of the total concentration of AO, the results shown in Fig. 52 support the increase of free AO with the decrease of R/D ratio. When

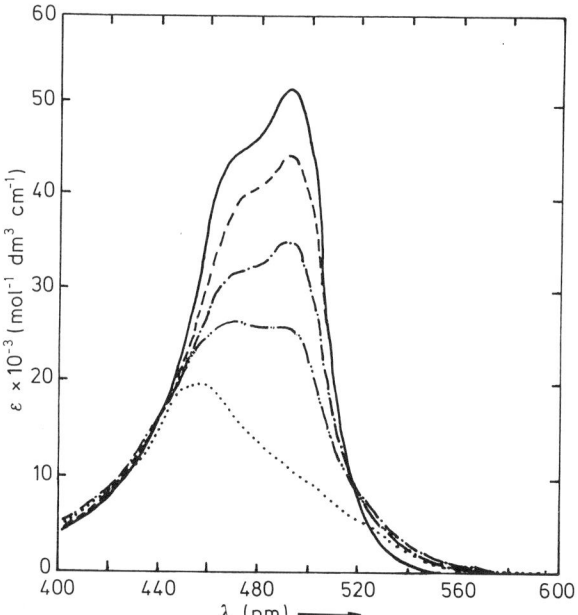

Fig. 52. R/D dependence of absorption spectra of PLGA-AO system at pH 4.5 [149].
R/D = 0: ─────
R/D = 1: ─ ─ ─ ─
R/D = 2: ─·─·─·
R/D = 4: ─··─··─
R/D = 10: ------
[AO] = 2×10^{-5} M

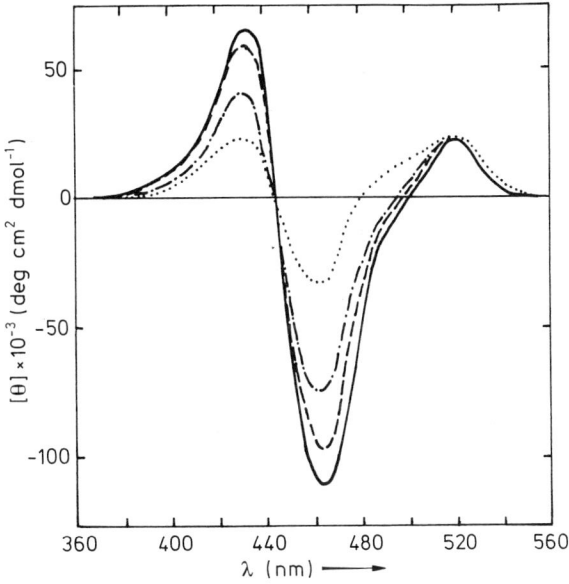

Fig. 53. R/D dependence of CD spectra of the PLGA-AO system at pH 4.5 [149].
R/D = 10: ———
R/D = 8: – – – –
R/D = 6: –·–·–·
R/D = 4: · · · · · ·
[AO] = 2×10^{-5} M

R/D is less than 10, the magnitude of the CD band at 222 nm decreased proportionally to that of the molar ellipticity at 210 nm. These results imply that the bound AO molecules lower the stability of the helical conformation of PLGA with a decreasing R/D ratio. When the pH is raised, PLGA transforms from a helix to a random coil. The effect of the addition of AO to the PLGA solution on the helix-coil transition of PLGA was studied at various R/D ratios. When the R/D ratio is large, the bound AO does not influence the pH dependence of the helix-coil transition. However, when the R/D ratio is smaller than 10, the pH region in which the helix-coil transition takes place is shifted upward. Cationic AO come to bind to the carboxylate anion, and contribute to the stabilization of the helical structure of PLGA. At a R/D ratio around unity, a characteristic variation curve was obtained, implying the helix formation in the neutral and alkaline pH region (Fig. 54). The array of the bound AO around the PLGA stabilizes the helix PLGA together with the AO array itself. Two ICD bands at 420 and 470 nm with a shoulder appeared in the PLGA-AO complex at an R/D of unity in the region of pH 8.3—10.3 and this may be due an aggregation of AO around PLGA (Fig. 55). Furthermore, the residue ellipticity of the CD band at 222 nm is extraordinarily large ($[\theta]_{222} = -1.2 \times 10^5$), compared to the value of $[\theta]_{222}$ expected in the helical PLGA. The orientation of the transition moments of AO molecules bound to the core of the right-handed α helix of PLGA results in an alignment of couplets, which themselves form a helical structure. It was assumed that the extraordinarily high $[\theta]_{222}$ value originates from the coupling of the spin-forbidden $n-\pi^*$

transition of polypeptides with any other allowed π–π* transition of AO. A similar observation was made in the complex of poly(L-lysine) (PLL) with hemin [150] or methyl orange (MO) [151] in alkaline aqueous solution. A group of strong bands near 500 to 550 nm indicates that a type of polymer aggregate of the AO molecules formed. Thus, it was assumed that the π–π* transition of AO molecules in the complex, as a polymer aggregate, was coupled mutually to produce strong CD bands in the region of 400–600 nm.

In the acidic pH region, it was found that one AO molecule interacts with three glutamyl residues in PLGA using a variation method [152,153]. Since AO molecules

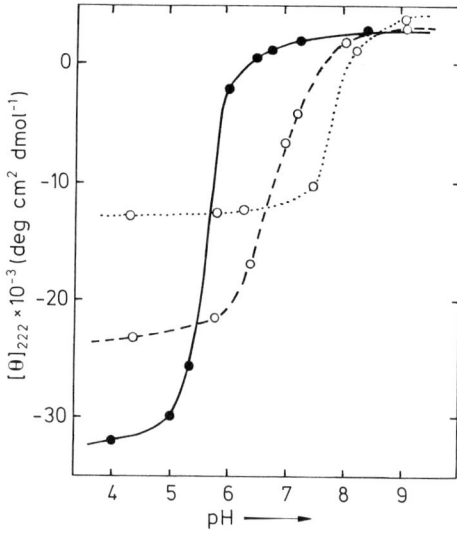

Fig. 54. Helix-coil transition of PLGA at various R/D ratios [149].
R/D = 10: ---○---
R/D = 4: ···○···
PLGA only: ——●——
[AO] = 2×10^{-5} M.

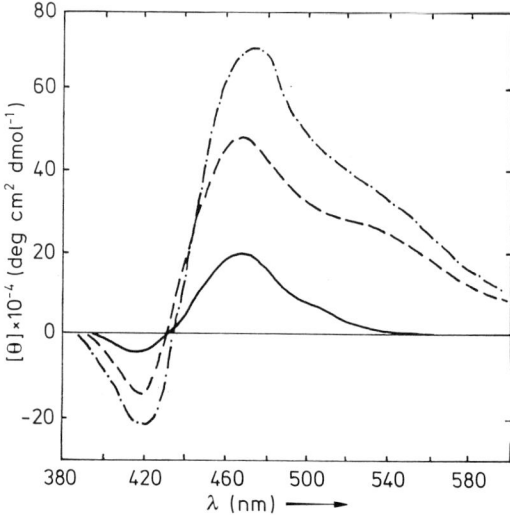

Fig. 55. The pH dependence of CD spectra of the PLGA-AO system at R/D = 1 [149].
pH = 8.3: ———
pH = 9.1: —·—·—
pH = 10.3: ————
[AO] = 2×10^{-5} M.

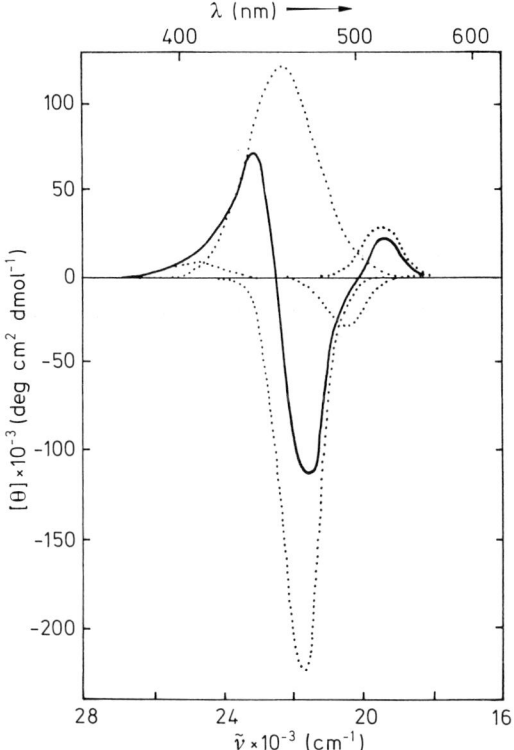

Fig. 56. CD spectrum of the PLGA-AO system and its resolution into Gaussian bands [153]; R/D = 10, pH = 4.5, [AO] = 2×10^{-5} M.
Observed spectrum: ———
Gaussian bands: - - - - - -

have a strong stacking tendency, the molecules are not randomly distributed among the available sites on a helical PLGA, but occupy sites adjacent to one another on the PLGA and stack with each other. A Gaussian analysis, as shown in Fig. 56, for the CD curve of PLGA at an acidic pH suggests the presence of one positive CD band at 520 nm and a negative CD band at 490 nm. The magnitudes of the two CD bands are nearly equal to each other, but the signs are opposite. The geometrical factor is assumed to be +1 because of an antiparallel sandwich-type AO dimer, and the distance from the center of one AO molecule to that of the nearest neighbot AO molecule can be estimated to be 7.6 Å, which is a reasonable value. The rotatory strength was estimated to be 25×10^{-40} c.g.s. if the AO molecule plane lies at an angle of 15° to the axis of the helix. The experimental value of the rotatory strength obtained from the CD magnitudes of the bands at 520 and 490 nm was 23×10^{-40} c.g.s. and obtained by the following equation:

$$R = 0.696 \times 10^{-42} \, \pi [\theta]_k \frac{\Delta_k}{\lambda_k}, \tag{78}$$

where $[\theta]_k$, λ_k, Δ_k are the molar ellipticity, the wavelength, and the half-band width of the k-th CD band, respectively. The aggregated AO molecules have a form of a left-handed superhelix around the core of the α helix of PLGA.

The complexes of poly(S-carboxymethyl-L-cysteine)-AO [154] and PLL-azo dyes [151, 154-156] have been analyzed in terms of ICD. MO is a derivative of azo-benzene substituted at 4,4'-positions with dimethylamino and sulfonate groups. In the presence of PLL, the absorption of MO is strongly dependent on the molar concentration ratio of the lysyl residue to MO(R/D) in the pH range of 3–10. The MO molecule does not stack, even when the concentration of MO is increased without PLL in a solution. However, the appearance of a new absorption band at 368 nm caused by the presence of PLL suggests that a dipole-dipole interaction of the transition moments of dimeric MO molecules bound to PLL. From a continuous variation method, the complex formation seems to be determined by the following factors: an electrostatic attraction, i.e., an ion pair formation between the lysyl cation and MO anions, and van der Waals' interaction. At the wavelength corresponding to the new absorption (368 nm), there are two CD bands which are nearly equal in magnitude but opposite in sign. The positive CD band appears at the shorter wavelength side. The new absorption band and these CD bands rapidly disappear when neutral salt is

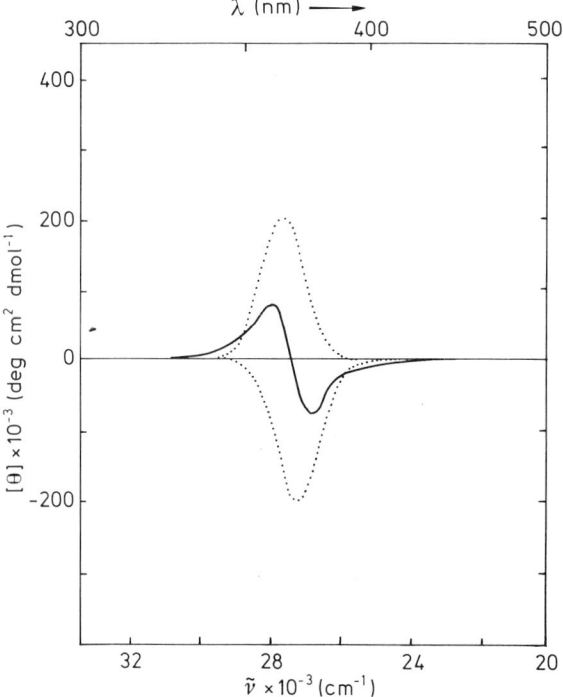

Fig. 57. CD spectrum of the PLL-MO system and its resolution into Gaussian bands [151]. R/D = 2, pH = 7.0, [MO] = 2×10^{-5} M.
Observed spectrum: ———
Gaussian bands: - - - - - -

added to the system. The CD bands were resolved into their components by a conventional Gaussian fitting method (Fig. 57) [151]. This indicates that a couple of CD bands is a typical exciton coupling arising from the dimeric MO bound to the lysyl residue in PLL which may be in random coil. The dimeric MO molecules, so-called skew dimers, are seen as chiral, and the chirality comes from the asymmetric α carbon atom of the lysyl residue in PLL. The magnitudes and signs of the CD bands show that S chirality is applicable to MO molecules bound to PLL. When poly(α,γ-diaminobutyric acid) was used instead of PLL, the signs of a couple of CD bands were inverted. The poly(L-arginine)MO complex also exhibits similar CD near 380 nm to that in PLL-MO [157] along with a positive CD band at 390 nm and a negative CD band at 374 nm below pH 9.6, while it showed a negative CD band at 374 nm and a positive CD band at 361 nm above pH 9.6. Thus, the interacting transition dipole moments of the dimeric dye can be assumed to have R chirality below pH 9.6 and S chirality above pH 9.6.

Recently, this ICD technique has been widely used for analyzing the interaction mode between a protein and a substrate and stoichiometry of the substrate bound to a protein. The hepatic microsomal monooxygenase system containing cytochrome P-450 catalyzes the oxidative metabolism of polycyclic hydrocarbons. It is well known that liver microsomes contain multiple molecular forms of cytochrome P-450 differing from one another in molecular and catalytic properties as well as inducibility by treatment in vivo with various drugs having aromaticity. Binding of various hydrocarbons to P-448$_1$ (the major cytochrome P-450 component of liver microsomes from methylcholanthrene-treated rabbits) induces characteristic CD bands in the wavelength region where the hydrocarbons themselves show absorption bands [158]. For example, pyrene exhibits absorption bands at 335, 320, and 305 nm and fluorescence emission bands at 375, 390, and 415 nm with excitation at 335 nm. When pyrene was mixed with P-448$_1$, negative CD peaks at 338, 323, and 308 nm appeared and corresponding quenching of the fluorescence emission was observed. As shown in Fig. 58, the CD magnitude at 338 nm increased as a function of the amount of added pyrene. The CD magnitude reached a plateau when an equimolar amount of pyrene was added to the cytochrome P-448$_1$ solution. The CD measurements were examined for 21 different hydrocarbons having 2 to 5 benzene rings. The hydrocarbons bind to P-448$_1$ at the same site and compete with one another [159].

We also have used ^{19}F NMR and ICD spectra to study interactions of trifluoperazine with calmodulin under various conditions where no aggregation of trifluoperazine occurs. Antipsychotic drugs such as trifluoperazine (TFP) are strongly bound to calmodulin (CaM) in the presence of Ca^{2+} and behave as potent CaM antagonists [160,161].

TFP itself is achiral and thus no CD band was observable at 254 and 303 nm of its absorption peaks. CaM (50 μM)-TFP (100 μM) solution in the presence of ethyleneglycol bis (β-aminoethylether)-N,N,N',N'-tetraacetic acid (EGTA) (2 mM) exhibits a large positive sharp CD band at 265 nm. By adding 0.2 M KCl to the CaM (50 μM)-TFP (100 μM) solution, a positive CD band at 265 nm disappears. Adding excess CaCl$_2$ (4 mM) to the solution caused a quite large CD trough around 260 nm together wich peaks around 293 and 315 nm and a trough around 360 nm. The CD trough at 260 nm was time-dependently enlarged and reached a maximum about 50 h after sample preparation. The CD bands will be induced by a chiral conformation

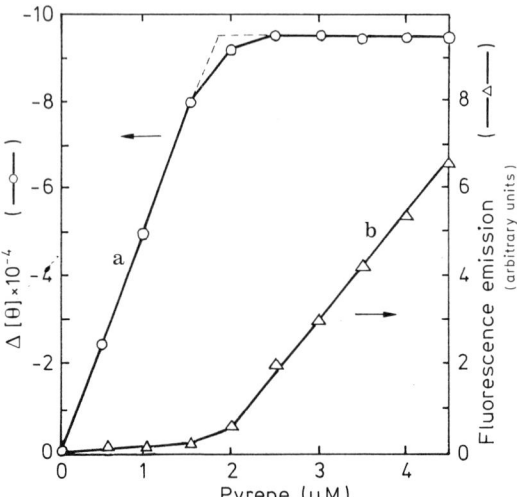

Fig. 58a and b. Titration of P-448₁ with pyrene [159]. The indicated amounts of pyrene were added to 1.92 μM P-448₁ in 0.3 M potassium phosphate buffer, pH 7.25, containing 20% glycerol. CD spectra and fluorescence emission spectra with excitation at 335 nm were measured.
a. Change in CD magnitude at 338 nm.
b. Fluorescence emission intensity at 375 nm

or configuration of a hydrophobic part of CaM with which TFP is interacting. From the time-dependent change of the ICD band at 262 nm, it was suggested that the CD change is composed of two phases, a fast phase with a half-saturation time of take out and close up 15 min and a slow phase with a half-saturation time around 25 h. In the presence of Ca^{2+}, the stoichiometry of TFP is 2 mol/mol of CaM. The TFP binding sites on CaM are classified into two groups; high- and low-affinity sites. The very slow conformational change in Ca^{2+}-CaM may be induced by the TFP binding, probably by the second TFP binding to the lower affinity site of Ca^{2+}-CaM [118]. The ICD of TFP bound to CaM is useful for analyzing the interaction mode of TFP with CaM in the presence of various kinds of metal cations [120]. Especially, the estimation of binding constants of TFP with CaM under various conditions can be successfully examined by ICD measurement. This estimation has been confirmed by NMR observations of various kinds of metal-ion nuclei, which are not conflicting with ^{19}F NMR results [118-120].

5.6 Induced Circular Dichroism of Heme and Chlorophyll Bound to Proteins

The ICD method has a high potential for research on allosteric effects in hemoglobin and the binding mode of chlorophyll and its derivatives to proteins and membranes. Hsu and Woody [193] investigated theoretically the origin of the ICD of heme in myoglobin and hemoglobin. The rotational strengths of the heme π–π* transitions in these two proteins [Q, B (Soret), N, and L bands] were calculated according to Kirkwood's theory [33] (see Sect. 3.3) as extended by Tinoco, Jr. [46,194], using atomic coordinates obtained by crystallographical analysis of Kendrew, Watson, and Perutz [195]. This calculation indicates that the coupled oscillator interaction between the heme transitions and allowed π–π* transitions in nearby aromatic side chains in the heme pocket within 2 Å induces the ICD bands in the wavelength regions of absorption of the heme. The calculated values of the rotational strengths are 0.3 and 0.1 DBM (Debye Bohr Magneton; 1 DBM = 0.9273×10^{-38} c.g.c. units) for the Soret bands of myoglobin and hemoglobin, respectively, and these values are quite similar to those obtained experimentally (0.5 and 0.2 DBM) for the two proteins. The calculation suggests that the contributions of the amide groups in the main chain to the ICD of heme bands can cancel each other out. This assumption is conceivable, because the two proteins are globular in shape and the contribution of nearby alkyl side chains to the ICD is likely to be small and not a dominant factor. The rotational strength is inversely proportional to the square of the distance between the heme and an aromatic group, assuming a definite relative orientation between these two groups [26]. Dramatic changes in the shape of the Soret CD band appear to follow the discussion of the binding mode of ligands to the heme at its apical position if the ligand binding does not produce significant reorientation of the heme relative to the protein moiety and the major conformational change in the protein. Woody's proposal has been applied to the ICD in cytochrome c peptide systems [196–198] and in simple model compounds for myoglobin and hemoglobin (Hemins I and II) [199].

Hemin I   Hemin II

When one axial ligand is the imidazole of histidine in a peptide or protein, the ICD depends on the nature of the sixth ligand group; the large Soret ICD for the weak ligand field groups such as H_2O and the small ICD for the strong ligand field groups such as imidazole [163–164]. These phenomena are attributed to the reduced symmetry of the heme group [200]. The high-spin species of the heme ligated with the weak ligand has an appreciable doming on the heme to reduce the symmetry around the heme iron.

On the other hand, the simple complexes (hemins I and II) have a reverse relation between the ICD magnitudes in the Soret region and the ligand field strengths of the axial ligands bound to the heme for peptides and proteins. This difference between our models [199] and Urry's findings [196-198] may be partly due to the existence of the ring(s) with optically active center(s).

An alternative example has been reported by Boxer et al. [201,202]. They formed well-defined complexes of chlorophyll and bacteriochlorophyllide with apomyoglobin. The chlorophyllide and its derivatives substitute for heme in the hydrophobic

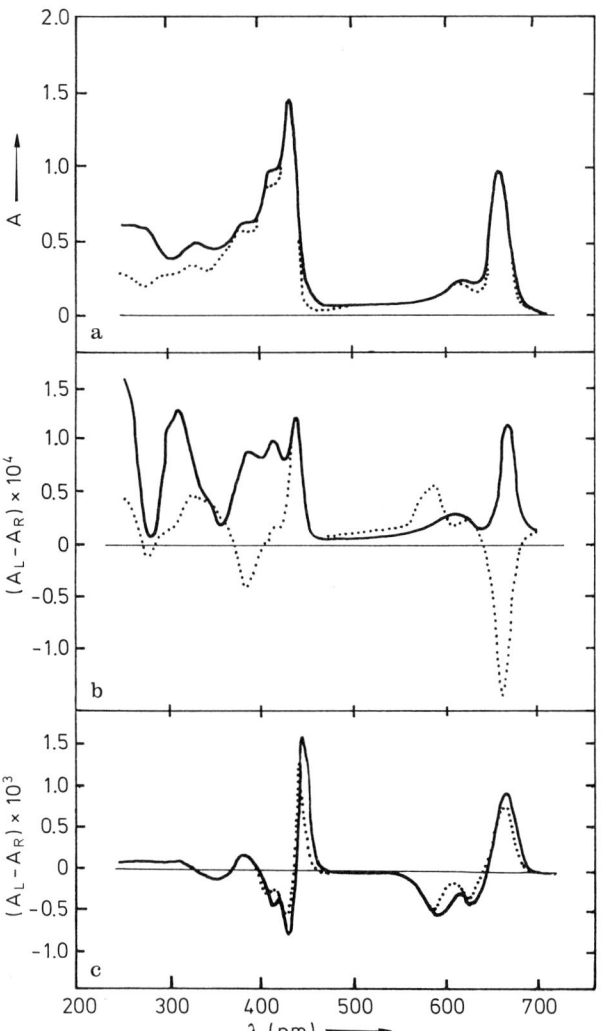

Fig. 59a–c. Absorption (**a**), CD (**b**), and MCD (**c**) of synthetic chlorophyll-apomyoglobin complexes [201].
Solid line: pyrochlorophyllide Mg(II)-apomyoglobin complex, 10^{-5} M in H_2O.
Dotted line: 3-(1-imidazolyl)propylaminated pyrochlorophyllide a Mg(II)-apomyoglobin complex, 10^{-5} M in CH_2Cl_2

pocket of myoglobin. The effect of the apomyoglobin on absorption, CD, MCD, NMR, and fluorescence lifetime in solution was compared with that for appropriate models in organic solvents. NMR provides valuable information on the structure and integrity of the chlorophyllide-apomyoglobin complexes, because the NMR chemical shift is sensitive to the location of the macrocycle ring current. Thus, each amino acid residue near the chlorophyll in the pocket of myoglobin was estimated by NMR chemical shifts. The chlorophylls have asymmetry centers themselves and well characterized intrinsic CD [201]. Figures 59 and 60 show that the CD spectra are substan-

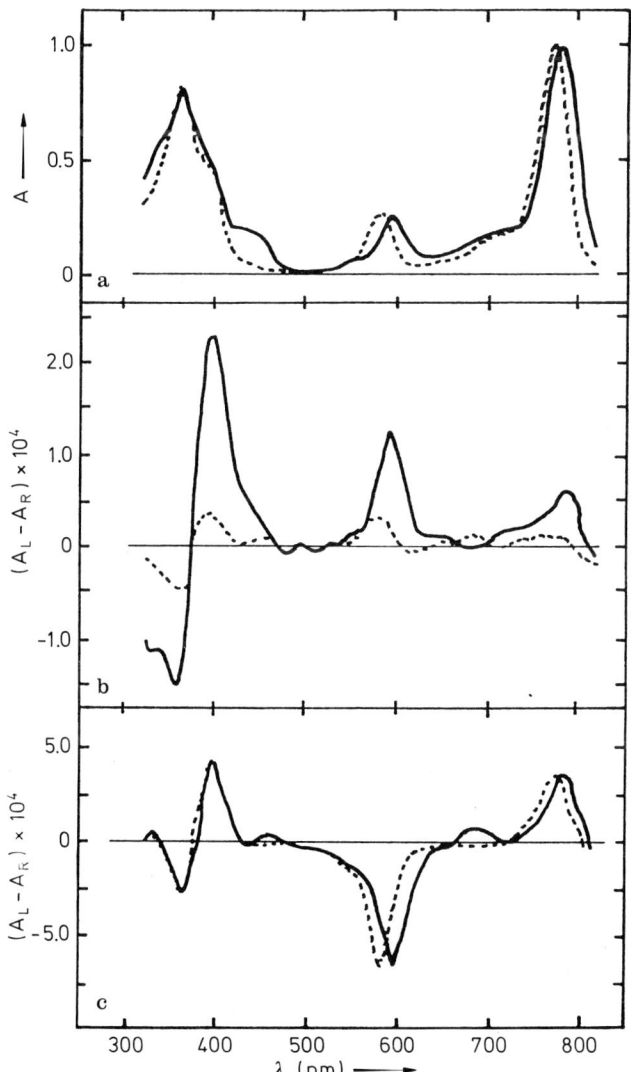

Fig. 60a–c. Absorption (**a**), CD (**b**), and MCD (**c**) of synthetic chlorophyll-apomyoglobin complexes [201].
Solid line: bacteriopyrochlorophyllide a Zn(II)-apomyoglobin complex, 10^{-5} M in H_2O.
Dotted line: bactriopyrochlorophyllide a Zn(II), 10^{-5} M in 1% pyridine in CH_2Cl_2

tially altered when all chlorophylls are inserted into apomyoglobin, while MCD spectra are relatively insensitive to the presence of the protein, as seen in Fig. 59c and 60c. The CD signs of several transitions are entirely reversed in the chlorines, and 3- to 6-fold increases are observed with zinc-bacterichlorophyllide a. But, no or less shifts are observed for both absorption and CD spectra. Thus, one can not ignore the effects of the protein on the ICD of chlorophyllides incorporated into the protein in shape and sign. To extend this approach to the analysis of allosteric effects in hemoglobin, zinc-pyrochlorophyllide a (ZnPchl a) was substituted for all four hemes and the two complementary hybrids in which ZnPchl a is substituted for heme in either the α or β chains, the heme remaining in the other chain, were synthesized [202].

The CD features of bacteriochlorophyll a in light-harvesting bacteriochlorophyll-protein complexes from native bacteria is more interesting and is not compatible with Boxer's experiments. Bacteriochlorophyll(Bchl)-protein complexes exhibit various CD spectral profiles, depending on the species of bacteria and their culture conditions [201-206]. Thus, CD of Bchl arises from the asymmetric environment in which the Bchl is situated, and from the specialized arrangement which affords a specific interaction between Bchl molecules, as well as from the asymmetry of Bchl

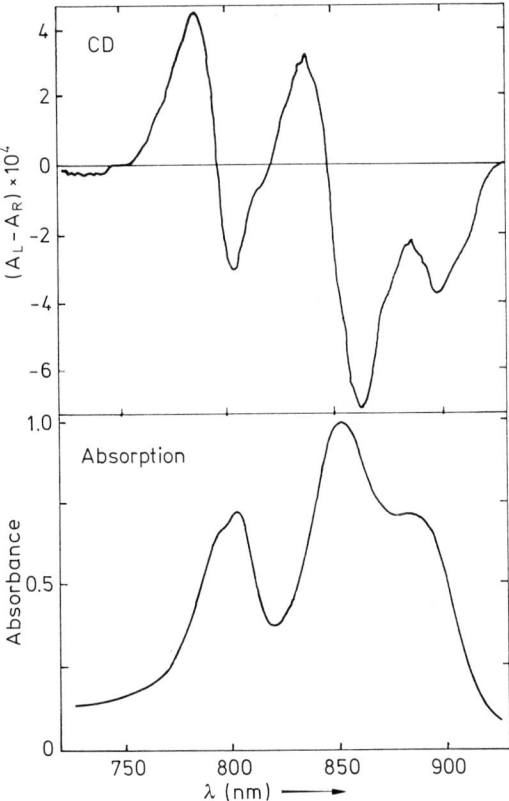

Fig. 61. Absorption (below) and CD (above) spectra of high B850 chromatophores from *Chromatium vinosum* [207]. It is noted that monomeric Bchl exhibits a sharp absorption peak at 770–780 nm

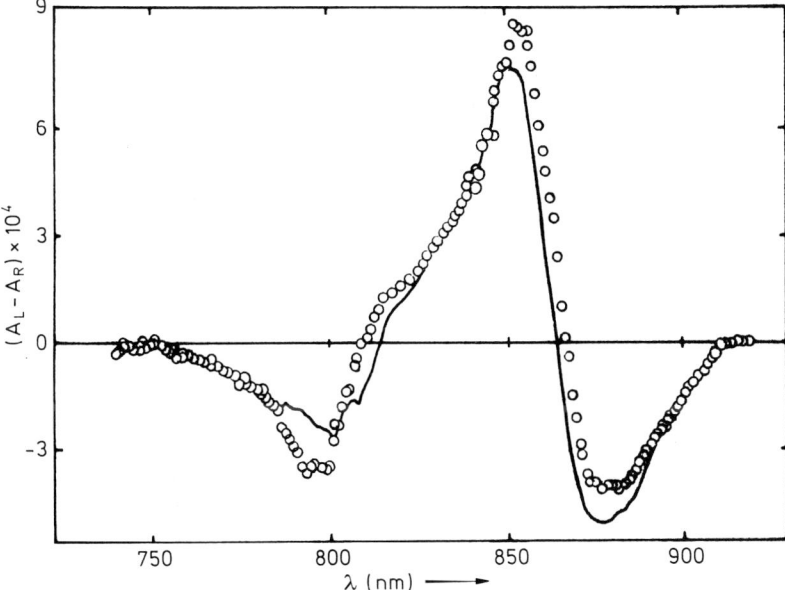

Fig. 62. Calculated CD spectrum from the CD spectra of isolated Bchl-protein complexes. The calculated CD spectrum (solid curve) was the sum of the CD spectra of B 870-reaction center complexes and B800–850 complexes from high 850 intracytoplasmic membrane in the proportion determined by the least-squares method to give the best fit to the observed CD spectrum (open circles) of high 850 intracytoplasmic membrane [208]

itself, when Bchl is buried in the protein(s). As shown in Fig. 61, absorption and CD spectra of chromatophores are shifted to an extraordinarily longer wavelength than those of the Bchl a in diethyl ether solution. We separated Bchl-protein complexes from *Chromatium vinosum* [207] and *Rhodopseudomonas palustris* [208]. These isolated Bchl-protein complexes retained the CD signals of light-harvesting Bchl samples which were observed *in situ* in chromatophores. Even if the samples including Bchl in intracytoplasmic membranes (ICM) exhibit nearly identical absorption, their states in ICM are not identical in terms of CD. Each Bchl sample was separated by polyacrylamide gel electrophoresis in buffer from the ICM of the bacterium, and the CD spectrum of each fraction was observed as a standard for the following calculation. Figure 62 shows the calculated CD spectrum (solid curve), which is the sum of the CD spectra of B870-RC (reaction center) and B800–850 complexes, with the observed CD spectrum of a sample of high 850-ICM (open circles). In this way, the amount of each component was estimated quantitatively. Bchl in the monomeric state exhibits CD peaks or troughs in the short wavelength region (<750 nm), while the Bchl-protein complexes extend the CD spectrum trace to longer wavelengths (>950 nm) depending on the bacterium culture and species. This large shift is not conceivable. Solvent variation can not shift the absorption and CD spectra beyond 80 nm [207] using Bchl in its monomeric state. Furthermore, doubly linked cofacial chlorophyll (Chl) dimer, hinged dimer of Chl, and singly linked open or folded dimer (Fig. 63) also did not shift the absorption and CD bands to longer wavelengths than 750 nm (see Fig. 64).

Thus, it is evident that neither of the solvent effect nor the effect of the stacking

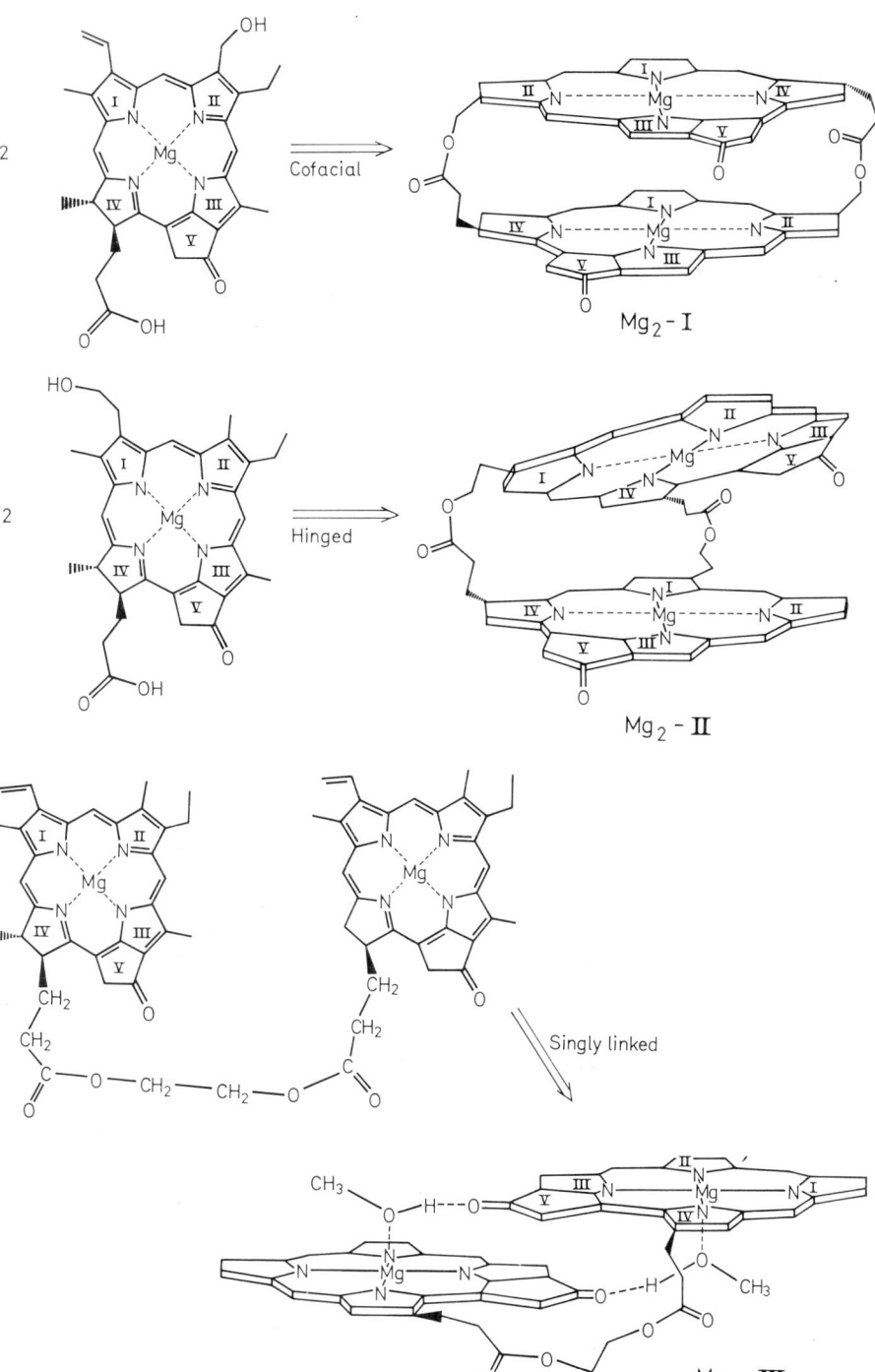

Fig. 63. Schematic diagrams of doubly linked cofacial dimer, hinged dimer, and singly linked open or folded dimer Chls [214]

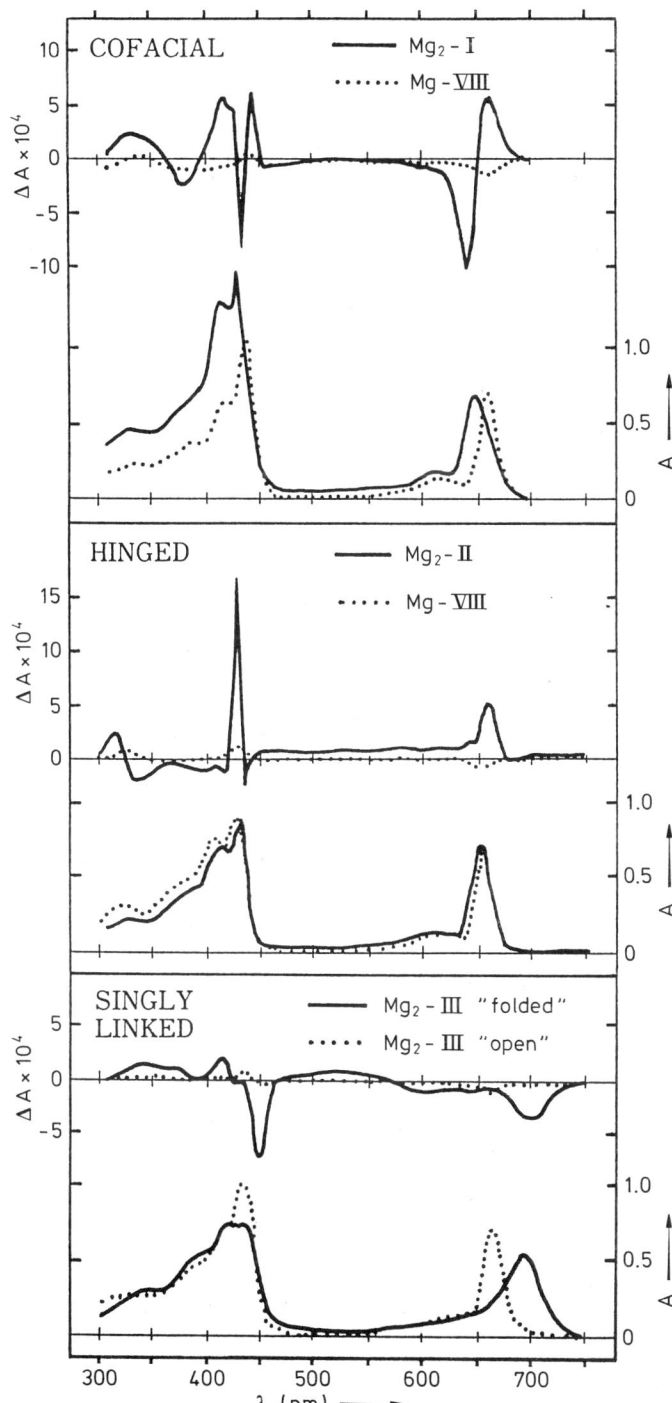

Fig. 64. Absorption and CD spectra of Chl-dimers (Mg$_2$-I, Mg$_2$-II, and Mg$_2$-III; see Fig. 63) and the corresponding monomer (Mg-VIII). The spectra of Mg$_2$-I, Mg$_2$-II, and Mg-VIII were obtained in CH$_2$Cl$_2$ with 0.5 M ethanol. The spectra of the "folded" singly linked dimer were obtained in toluene with 0.5 M ethanol, and the open form is the same sample after pyridine was added (10% by volume) [214)]

mode on the extraordinarily long shift of the CD is plausible for those of the Bchl-protein complexes.

Recently, the amino acid sequences of the light-harvesting Bchl-protein have been determined by direct protein sequence analyses [209-211] and DNA sequences of the encoding genes [212-214]. Interspecies comparison of these sequences revealed characteristic features among them. The light-harvesting Bchl-proteins which absorb around 870 and 890 nm are called B870 or B890 complexes, respectively. Nozawa and Hatano [191] discussed the interspecies comparison of B870 and B890 Bchl-proteins. The B870 α-polypeptides of *Rhodopseudomonas sphaeroides* and of *Rhodospirillum rubrum* show an essential homology in their amino acid sequences. In these sequences it is characteristic that the histidine residue, which is a potential candidate for the Bchl binding, is surrounded by hydrophobic amino acids. A similar disposition can be found from the B870 β-polypeptide of *Rhodopseudomonas capsulata*. According to the Kyte and Doolittle method [186], the distribution of hydrophobic residues along the polypeptide chains indicates one membrane-spanning region for each B870 polypeptide. The histidine residue exists close to the boundary in the hydrophobic segment. Presumably, the Bchls are linked to these chains one by one with the coordination bond between the imidazole nitrogen of the histidine residue and the Mg(II) ion of Bchl. The picture is consistent with the resonance Raman results which

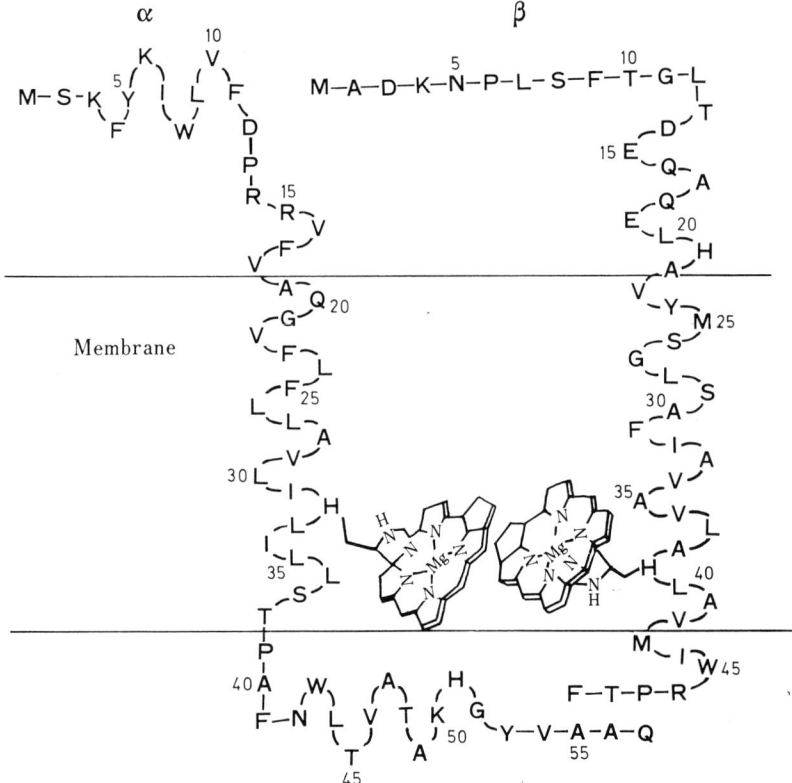

Fig. 65. A schematic model for the light-harvesting (LH1) complex of *Rhodopseudomonas capsulata* [191]

indicate the presence of five coordinated Bchls [216], and also with the CD results which indicate strong interactions between the Bchls [205,207,208]. A schematic model for the B870-complex from *Rhodopseudomonas capsulata* is shown in Fig. 65. The hydrophobic segment is tightly bound, together with the Bchl molecule, to the membrane, and this opinion was confirmed by ^{13}C NMR of the B870-complex with membrane by the cross polarization/magic angle spinning technique [216].

5.7 Metal Binding to Proteins

Specific interactions between metal ions and proteins serve numerous important biochemical and biological functions in areas as diverse as enzymatic reactions, muscle contraction, cell motility, neurotransmission, and hormonal regulation of metabolic pathways [217]. Ca^{2+} and Zn^{2+} ions are of much interest in biological systems. Certain enzymes require Ca^{2+} or Zn^{2+} ions as a cofactor, the cations being tightly bound to or near the active site. Others are merely stabilized by Ca^{2+}, Zn^{2+}, or Mg^{2+} bound at noncatalytic sites. For example, Ca^{2+} is bound to the active sites of phospholipase A_2 [218] and staphylococcal nuclease [219] for their activation. Thermoparvalbumin [222] and troponin C [221] exhibit characteristic conformational changes to form α-helical segments upon adding stoichiometric amounts of Ca^{2+} to the proteins. The third class of Ca^{2+}-binding proteins has both high- and low-affinity sites, and the quantitative estimation of the binding constants of Ca^{2+} to the sites is necessary for understanding their functions.

Ca^{2+} and Zn^{2+} ions are diamagnetic, and colorless in the visible region. The unique value of the Co^{2+} ion lies in the fact that it is paramagnetic and has distinct optical and magnetic circular dichroism (MCD) properties. Electron spin resonance and MCD data provide useful information on the effective symmetry around the paramagnetic metal ion bound to the binding sites on the proteins. Especially the MCD technique is quite powerful for the detection or differentiation of the effective symmetry around the paramagnetic probe (see Appendix). One should pay much attention to selecting the proper substituting metal ions. For example, Co^{2+} only can activate the Zn^{2+}-depleted D-lactate dehydrogenase, whereas other metal ions such as Mg^{2+}, Mn^{2+}, Cd^{2+}, or Cu^{2+} did not cause any enzymatic activity or any spectral change when added to the Zn^{2+}-free enzymes [224]. The same situation has been found for Tb^{3+} addition to calmodulin. another example has been indicated for *E. coli* alkaline phosphatase [225]. Figure 66 shows the similarity of the ICD from the fixed aromatic side chains in the protein to the Co^{2+} ion to that in the protein bound to Zn^{2+}. CD titration reveals that only 0.5 g-atom of Zn^{2+} or Co^{2+}/monomer of the apoenzyme is required to evoke the maximal spectral change. In the titration of the apoenzyme with Co^{2+}, the presence of Mg^{2+} neither affects the stoichiometry nor final CD spectrum. The effect of other coexisting metal ions on the binding of a given metal ion to a protein can be analyzed in more detail by the observation of change in the line-widths of the NMR spectrum of each metal-ion nucleus [120,226]. In these experiments, the titration curves, as indicated by the CD and absorptivity, is substantially important for the analysis of metal reconstitution of an apoenzyme. Holmquist et al. have explored the apparent nonequivalence between the two active-site structures in yeast superoxide dismutase by spectral analysis of the binding sites during Co^{2+} titration of the apo-superoxide dismutase [227].

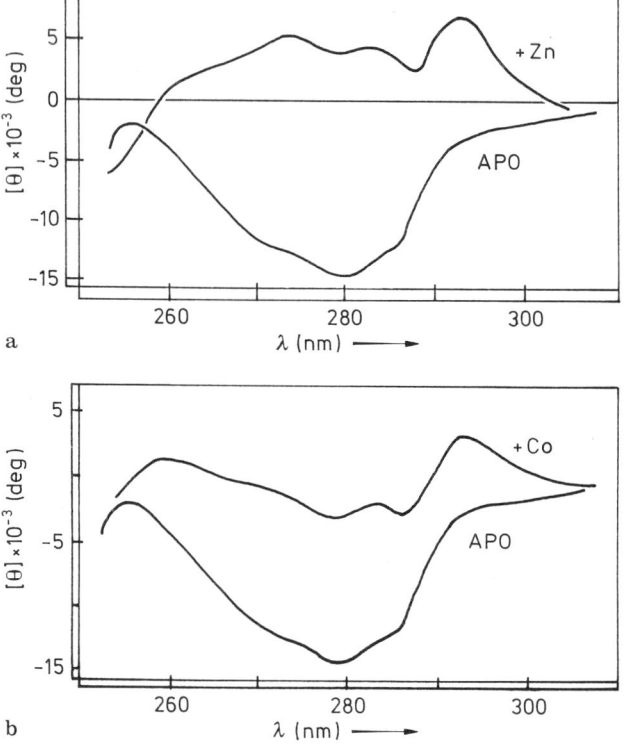

Fig. 66a and b. CD spectra of Zn(II) (**a**) and Co(II) (**b**) complexes with *E. coli* alkaline phosphatase in the monomeric state, compared with the CD of its apoenzyme [225]

5.8 Future Trends and Scope on Induced Circular Dichroism in Protein-Dye Systems

There are many chromophores in a typical protein. The three aromatic amino acids phenylalanine, tryptophan, and tyrosine, as well as the disulfide linkage between cysteine residues are present in most proteins, and they become the origin for the near ultraviolet CD bands. The intensity and positions of the CD bands depend strongly on the rigidity in space of the aromatic CD as well as disulfide CD often fluctuate in their position and magnitudes. The complexity makes it very difficult to obtain any definite structural information from the CD of the proteins in this region. So, one should use complementarily other physicochemical tools such as NMR or fluorescence spectroscopy. For example, histidine and tyrosine residues exhibit fairly separated chemical shifts of ^{1}H and ^{13}C NMR signals from other side-chain peaks, and the NMR signals of the amino acids are very sensitive to ionization. Further, T_1 and T_2 values of these resonances can be connected with their rigidity in space [216]. With the development of instruments capable of high-precision measurements of CD, it has become possible to study small changes in the CD spectrum of proteins, reflecting the protein conformation change, upon binding substrate, inhibitor, coenzyme or metal ions. Furthermore, rapid observation of the CD change is appreciable for analyzing the conformational change of a protein and for determining the reaction rate of a reaction between a protein and added reactants, using a stopped-flow system [228-231] or temperature-jump system [232]. Further, high utility of rapid scanning spectropolarimeters, using an acoustic optical filter, has been attractive as of its facility [233].

6 Induced Circular Dichroism in Polysaccharide-Dye Systems

The solid-state structures of polysaccharides have been characterized by X-ray diffraction, ^{13}C NMR measured by cross polarization/magic angle spinning method, and infrared spectroscopy. Polysaccharides, however, may have various types of conformation in solution. Conformational changes and anomeric changes in solution are often connected with their physiological activities. Previously, only the ORD of polysaccharides could generally be measured, and CD studies were possible in those cases where other chromophores were present in their molecules. Alternatively, the ICD of bound dyes has been measured for analyzing conformational changes in solution. All monosaccharides, oligosaccharides, and polysaccharides have no electronic transition in wavelength regions longer than 190 nm. However, commercially available CD spectrometers are currently limited to measurements above approximately 190 nm, since atmospheric dioxygen begins to absorb very strongly at wavelengths shorter than 180 nm. It is now possible to measure the CD of polysaccharides directly, even when electronic transitions are present below 180 nm. For CD measurements in the region from 190 to 100 nm, we need to use an evacuated CD spectrometer equipped with a MgF_2 polarizer and CaF_2 quarter-wave retarder. At the present stage, it is too early to predict the exact nature of the conformational information of a family of sugars by vacuum ultraviolet (VUV) CD, and it is emphasized to utilize the ICD of the dyes bound to the sugars for predicting conformational changes and anomeric changes in solution.

6.1 Vacuum Ultraviolet Circular Dichroism and Its Applications to Saccharides

Some molecules of biological interest absorb only below 190 nm, where light absorption by dioxygen in air asserts technical limitations. This is overcome by evacuation of the entire CD spectrometer chamber and by using optical elements which have no absorption in the desired wavelength region. Successful VUV CD spectrometers have been described in the literature [234,235], several of which have used a 1 m focal length vacuum monochromator capable of moderately high resolution and wavelengths as short as 130 nm. For a spectrum of the molecule in solution, the wavelength range below 160 nm is not accessible owing to absorption by the solvent. Nelson and Johnson Jr. [234,235] reported the CD bands in the region of 167–180 nm of monosaccharides such as D-glucose, D-xylose, and D-galactose. Further, the CD bands at 185 nm of various methyl glucosides are assigned to the oxygen atom in their pyranose ring, and their CD bands at 175 and 165 nm to the oxygen atom in the methoxy group. The latter two invert in signs, correlating their stereochemistry around the anomeric carbon atoms. Stevens et al.[236] revealed that the CD bands in the regions of 180–190 nm, 164–177 nm, and 145–160 nm depend on the glycosidic linkages of α- and β-(1 → 3), (1 → 4), and (1 → 6) types, respectively, α-1,4-D-Glucan (amylose) and β-1,6-D-glucan, which can form a gel, exhibit a positive CD band in the region of 180–190 nm, while α-1,6-D-glucan (dextran), which has relatively high flexibility in its main chain, exhibits no CD in this region. All D-glucans show CD bands in the region of 164–177 nm, where the signs depend on the anomer orientation for 1 → 3 and 1 → 4 linkages. Thus, the positive CD appears for α type, while the negative for β type. On the other hand, the positive CD in this region can be expected to be due to a 1 → 6 linkage.

6.2 Dye Probes for Polysaccharide Conformation Analysis

Little has been known about ICD of dyes bound to polysaccharides. Nishida and Iwasaki[237] first found that AO molecules bound to carboxymethylcellulose (CMC) in aqueous solution exhibits ICD bands in its absorption region. The CD signs, positive at 425 nm and negative at 450 nm, were assigned to a left-handed configuration of the bound AO molecules. Hyaluronic acid also was analyzed in terms of the ICD of AO bound to it[238]. In these cases, the helical handness of the polysaccharides had not been well defined.

A bacterial (1 → 3)-β-D-glucan, curdlan, which entirely consists of β-(1 → 3)-linked D-glucose residues, is soluble in aqueous alkaline solutions but not in neutral or acidic solutions, and it forms a firm resilient gel when its aqueous suspension (at pH < 12) is heated above 54 °C or its alkaline solution is gently neutralized to below pH 12. The helical sense had not yet been determined, however, though X-ray diffraction data revealed that the chain conformation of this polysaccharide is a triple-stranded 6_1 helix with a right-handed sense. Ogawa et al.[239] revealed that the ordered structure of the polymer in a dilute alkaline solution is retained in a gel or in neutral solution. This conclusion came from the experimental results that no or less change in CD signs of the ICD of Congo Red bound to the D-glucan can be detected on changing the pH in the solution.

Since the observations of the ICD were on equilibrium mixtures between anomers, definitive interpretations must await for other physicochemical analyses such as NMR.

6.3 CD Analysis of Side-Chain Chromophores on Saccharides

The ICD of the dyes bound to saccharides through an ionic coupling or hydrophobic interaction may remain a conflicting problem. The side-chain chromophores covalently bound to saccharides permit CD bands in the far or near ultraviolet region. These side-chain chromophores can exhibit CD and thus provide more definitive information on the conformation of saccharide moieties. Thus, acetamide CD has been observed to reflect polymer secondary structure of glycosaminoglycans, which are the connective-tissue proteoglycans.

Glycosaminoglycan CD is complicated by the presence of two chromophoric groups with overlapping electronic transitions, the carboxyl and acetamide substituents.

In a comparison of the solution and solid-state CD of sodium hyaluronate, hyaluronic acid, and chondroitins [240-242] some CD bands were assigned to the amide groups. Stevens et al. [243] assigned the CD bands in the far and vacuum ultraviolet region of both chitan and chitin in solution, gel, and film. The samples were dissolved in hexafluoro-2-propanol (HFIP). Hydrolysis does not occur in HFIP as evidenced by the constancy of sodium D-line rotation of these solutions ($-58°$, c 0.1). The solution spectrum of chitin in HFIP (Fig. 67) is different from that of the related monomer methyl-2-acetamido-2-deoxy-β-D-glucopyranoside and dimer 2-acetamido-4-O-(2-acetamido-2-deoxy-β-D-glucopyranosyl)-2-deoxy-D-glucose (chitobiose) in HFIP. However, its spectrum is similar to that of 2-acetamido-2-deoxy-D-sugars, differing from chitin in configuration at C–4 or C–1 or in the linkage position. The CD spectrum of films cast from solutions of chitin in HFIP showed an increase of CD magnitudes and a sign change in the n–π* region (Fig. 67). Absorption spectra show a red shift from 186 nm in solution to 194 nm in film. This red shift leads to the assumption that the CD extremum at 197 nm of the film can be assigned to a π–π* transition. Figure 68 shows the CD spectrum of 2-acetamido-2-deoxy-D-glucosamine film cast from HFIP.

The CD spectra of chitan in HFIP solution and film are shown in Fig. 69. A comparison of chitin and chitan spectra shows that free amine groups of chitin do not affect the spectral profile of the CD. It is noted that a red shift of the π–π* absorption maximum accompanied film formation. The solid state structures of chitin and chitan have been characterized by X-ray and infrared spectroscopy. Three forms have been proposed: parallel, antiparallel, and mixed. Infrared analysis revealed that the amide groups are in *trans* conformation and the amide plane is nearly perpendicular to the ring plane of the amide oxygen eclipsing the C–2 hydrogen. The *trans* amide conformation may permit the amide groups to particitate in two types of hydrogen bondings as both donor and acceptor. The *trans* amide conformation and intermolecular hydrogen bonding are assumed to be restricted in the solid state as inferred from the CD in the solid state.

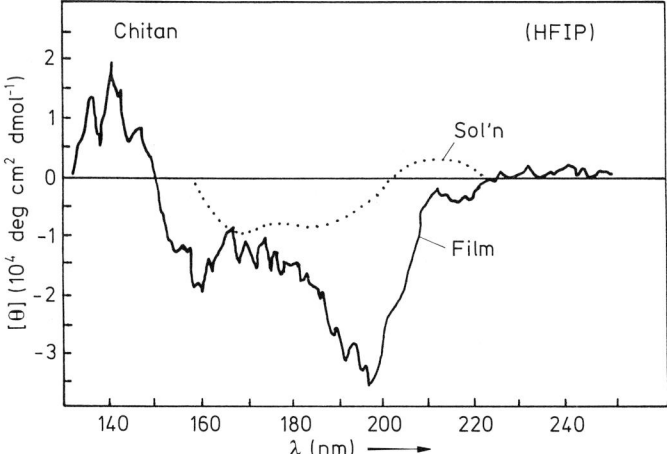

Fig. 67. CD spectra of chitin in HFIP solution (dashed line) and a chitan film cast from its HFIP solution (solid line) [243]

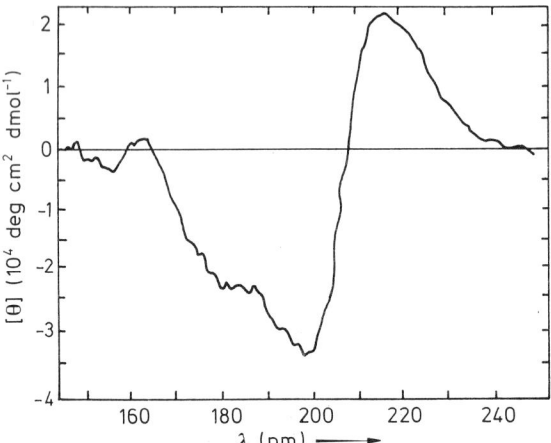

Fig. 68. CD spectrum of 2-acetamido-2-deoxy-D-glucosamine film cast from its HFIP solution [243]

Recent biochemical experiments revealing the extremely complex biosynthetic pathway to *N*-glycosylation of proteins suggest the biological importance of the carbohydrate moiety. Some insight into glycoprotein functions would be obtained from their structural analyses. Although ^{13}C NMR approaches to glycoproteins and glycopeptides provide exact and comprehensive information on their structures [244, 245], the CD approaches, especially those in the VUV region, will become more important for their conformational and anomeric analyses because of high sensitivity of the optical activity on their handness.

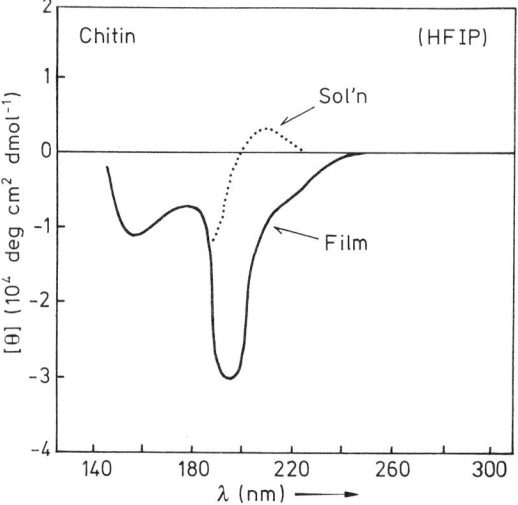

Fig. 69. CD spectra of chitan in HFIP solution (dashed line) and a chitan film cast from its HFIP solution (solid line) [243]

6.4 Benzoates Exciton Rule for Determining the Anomeric Configuration in Saccharides

The dibenzoate exciton rule has been widely applied for determining the absolute configurations of complex natural substances [26]. Liu and Nakanishi [246] extended the rule to 40 bis-, tri-, and tetrakis(p-bromobenzoate)s of various methyl pyranosides. The amplitudes, abbreviated as A, of split CD curves, i.e., the difference of the split extrema, increase, when the chiralities between the chromophores are all of identical signs. Otherwise, the magnitudes cancel each other out, and often show an uncharacteristic unsplit CD curve when opposing chiralities are present. Further, it was noted that an interesting additivity relation is present in the A values of the split CD curves of compounds containing three or more interacting chromophores. For example, the amplitude of the split CD curve of a molecule with three interacting chromophores can be approximated by the sum of the component amplitudes. The data for bis(p-bromobenzoate) derivatives are grouped into the following six classes:

Class I: 1,2 ee and 1,2 ea, $A = 62$ (in $\Delta\varepsilon$),
Class II: 1,2 aa, $A = 5$,
Class III: 1,3 ee, $A = 0$,
Class IV: 1,3 ea, $A = 16$,
Class V: Bis(benzoates) involving 6-OBz with 4(e) substituents,
  α-Glu-2,3-di-OMe-4(e),6- , $A = 15$
  α-Man-3(e),6- , $A = 11$
  α-Glu-2(e),6- , $A = -7$
  α-Man-2(a),6- , $A = 10$
Class VI: Bis(benzoate)s involving 6-OBz with 4(a) substituents,
  α-Gal-2,3-di-OMe-4(a),6 , $A = -15$
  β-Gal-3(e),6- , $A = -9$
  β-Gal-2(e),6- , $A = -4$.

In this table, the signs of A are that of the first or longer CD extremum, and the abbreviations used are as follows: Glu, Man, and Gal are glucose, mannose, and galactose, respectively. The A values, as mentioned above, reflect only the spatial arrangement of the benzoate groups, and are independent of other nonchromophoric substituents on the pyranose ring. This relation is applicable to the CD amplitudes of tris- and tetrakis(benzoate)s, approximated by adding the amplitudes of the components. This chirality rule should be applied to more complex molecules such as oligosaccharides and polysaccharides in terms of their handness.

6.5 Future Trends and Scope on Circular Dichroism in Saccharides

The CD of saccharides has been less documented than those of proteins and nucleic acids, because of the difficulty of the CD in the VUV region. Snyder et al. [247,248] reported the useful application of synchrotron orbital radiation (SOR) for VUV CD measurements. SOR from modern electron storage rings is highly linearly polarized and more intense than conventional VUV continuum sources. These properties make SOR ideal for CD measurements because of the better signal-to-noise ratio resulting from the higher light flux. The other advantages of SOR is that the circularly polarized light is easily available. In addition, SOR does not require the use of a MgF_2 polarizer which has limited the range of CD measurements to 135 nm. Use of the LiF quarter-wave retarder permits measurements up to 115 nm.

CD documentation may become more feasible by use of SOR for VUV CD measurements, which would be applied to more biological samples such as viruses and cell walls containing saccharides or glycoproteins.

7 Induced Circular Dichroism in Liquid Crystalline Phases

In Section 3.4.6 it was noted that chiral mesophases behave as chiral perturbation moieties for achiral molecules inserted to the mesophases. Here, only relations between the observed signs in LCICD and the pitch-dependent CD in liquid crystalline systems, and the helical sense of the liquid crystalline phases are described.

Several unique optical properties arise from helical structures of cholesteric liquid crystals.
1) For light incident parallel to the helix axis in a cholesteric phase, only the lowest-order reflection is allowed.
2) Only the component of optical polarization for which an instantaneous electric field matches the helical cholesteric directors is strongly reflected.
3) The reflection of only one component, RCP or LCP, in incident linearly polarized light can be observed as a very strong rotatory power.
4) The helical directors in a cholesteric phase can induce optical activity for an added dye intercalated into a cholesteric phase.

These unique optical properties of cholesteric liquid crystals have been investigated by de Vries [65].

7.1 The Relationship Between the CD Sign and the Helical Sense in Cholesteric Phases

A system consisting of helical directors reflects either RCP or LCP. Another component, LCP or RCP, in incident linearly polarized light is transmitted with no significant reflection loss. Thus, one can measure a CD band with a plus or minus sign when linearly polarized light passes through a cholesteric phase orientated to take its directors perpendicular to the light. Then, the sign of the observed CD depending on the cholesteric pitch provides the helical sense of a given cholesteric liquid crystal. However, this relation has been often erroneously inferred.

Now, one can consider a wave propagating along the helix axis z as shown in Fig. 2. The RCP is right-handed, moving along the positive z axis at one snap shot time. On viewing from the origin ($z = 0$) to the positive z axis, one can define the helix of directors as right-handed if the directors rotate clockwise.

From Maxwell's equation, the electric vector $\vec{E}$ interacting with a medium having dielectric constants parallel and perpendicular to the z axis, $\varepsilon_{\parallel}$ and $\varepsilon_{\perp}$, is described as:

$$\partial^2 \vec{E}/\partial z^2 = (2\pi i m \varepsilon_0^{1/2}/\lambda)^2 \, \vec{u} E_0 \exp[-i\{\omega t - (2\pi m \varepsilon_0^{1/2} z)/\lambda\}]$$
$$= -(4\pi^2 m^2 \varepsilon_0/\lambda^2) \, \vec{u} E_0 \exp[-i\{\omega t - (2\pi m \varepsilon_0^{1/2} z)/\lambda\}], \qquad (79)$$

Induced Circular Dichroism in Biopolymer-Dye Systems

where the director vector $\vec{u}$ arises from the anisotropy of the dielectric constants, $\varepsilon_\parallel$, and $\varepsilon_\perp$; $\varepsilon_0 = (\varepsilon_\parallel + \varepsilon_\perp)/2$; α is defined as $(\varepsilon_\parallel - \varepsilon_\perp)/(\varepsilon_\parallel + \varepsilon_\perp)$. E_0 is the amplitude of the wave at wavelength λ, which is propagating through the medium. The parameter m is a function depending on the cholesteric phase pitch P and on α. On transforming the coordinates to the rotating frame, one can utilize Pauli matrices as follows:

$$\sigma_1 = \begin{pmatrix} 0 & 1 \\ 1 & 0 \end{pmatrix}, \quad \sigma_2 = \begin{pmatrix} 0 & -i \\ i & 0 \end{pmatrix}, \quad \sigma_3 = \begin{pmatrix} 0 & 0 \\ 1 & -1 \end{pmatrix}, \tag{80}$$

and $\exp(-i\sigma_j \theta) = \sigma_j \cos\theta - i\sigma_j \sin\theta$,

$$\sigma_0 = \begin{pmatrix} 1 & 0 \\ 0 & 1 \end{pmatrix}, \tag{81}$$

$$\theta = 2\pi z/P.$$

The last equation is a relation between the pitch P and the rotating frame parameter θ. On using Pauli matrices, one can obtain the following equations from Eq. (79):

$$\partial^2 \vec{E}/\partial z^2 = -(4\pi^2 m^2 \varepsilon_0/\lambda^2) \sigma_0 \vec{u} \exp[-i\{\omega t - (2\pi m \varepsilon_0^{1/2} z)\lambda\}] \tag{82}$$

$$= -(\omega^2 \varepsilon_0 c^2)(4\pi^2 c^2/\omega^2 \varepsilon_0)(m^2 \varepsilon_0 \sigma_0/\lambda^2) \vec{u} \exp[-i\{\omega t - (2\pi m \varepsilon_0^{1/2} z)/\lambda\}] \tag{83}$$

and

$$-(4\pi i \sigma_2/P)(\partial \vec{E}/\partial z) = -(4\pi i \sigma_2/P)(2\pi i m \varepsilon_0^{1/2}/\lambda) \vec{u} \exp[-i\{\omega t - (2\pi m \varepsilon_0^{1/2} z)\lambda\}] \tag{84}$$

$$= (\omega^2 \varepsilon_0/c^2)(c^2/\omega^2 \varepsilon_0)(8\pi^2 m \varepsilon_0^{1/2} \sigma_2/P\lambda) \vec{u} \exp[-i\{\omega t - (2\pi m \varepsilon_0^{1/2} z)/\lambda\}]. \tag{85}$$

Again, from Maxwell's equation, one can obtain the following:

$$(\partial^2 \vec{E}/\partial z^2) + (\omega/c)^2 \vec{D}$$
$$= (\partial^2 \vec{E}/\partial z^2) - (4\pi i \sigma_2/P)(\partial \vec{E}/\partial z) + [(\omega/c)^2 (\varepsilon_0 \sigma_0 + \varepsilon_1 \sigma_3) - (4\pi^2 \sigma_0/P^2)] \vec{E} = 0. \tag{86}$$

If Eqs. (83) and (85) are inserted into Eq. (86), the following equation results:

$$\{-(4\pi^2 c^2/\omega^2 \varepsilon_0)(m^2 \varepsilon_0 \sigma_0/\lambda^2) + (4\pi^2 c^2/\omega^2 \varepsilon_0)(2m \varepsilon_0^{1/2} \sigma_2/P\lambda) + \sigma_0 + (\varepsilon_1/\varepsilon_0)\sigma_3 - (4\pi^2 c^2/\omega^2 \varepsilon_0)(\sigma_0/P)\} \times (\omega^2 \varepsilon_0/c^2) \vec{u} \exp[-i\{\omega t - (2\pi m \varepsilon_0^{1/2} z)/\lambda\}] = 0. \tag{87}$$

Here, $4\pi^2 c^2/\omega^2 = \lambda^2$, $\alpha = \varepsilon_1/\varepsilon_0$, and if $\lambda' = \lambda/\varepsilon_0^{1/2}$, we can transform to

$$[-(\lambda')^2 \sigma_0 - 2m\sigma_2 \lambda' + m^2 \sigma_0 \lambda' + \sigma_0 - \alpha \sigma_3] \times \vec{u} \exp[-i\{\omega t - (2\pi m \varepsilon_0^{1/2} z/\lambda'\}] = 0, \tag{88}$$

and finally to

$$\{[1 - (\lambda')^2 + m^2\lambda']\sigma_0 + \alpha\sigma_3 - 2m\lambda'\sigma_2\} \vec{u} \exp[-i\{\omega t - (2\pi m\varepsilon_0^{1/2}z)\lambda\}] = 0. \tag{89}$$

If

$$\vec{u} = \begin{pmatrix} \mu_1 \\ \mu_2 \end{pmatrix},$$

Eq. (89) is rewritte to:

$$\begin{bmatrix} 1 - (\lambda')^2 - m^2 + \alpha & -2im\lambda' \\ 2im\lambda' & 1 - (\lambda')^2 - m^2 - \alpha \end{bmatrix} \begin{bmatrix} \mu_1 \\ \mu_2 \end{bmatrix} = 0. \tag{90}$$

Accordingly, Eq. (91) is obtained, and finally reduced to Eq. (92):

$$m^4 - 2[1 + (\lambda')^2]m^2 + (\lambda')^2]^2 - \alpha^2 = 0 \tag{91}$$

$$m^2 = 1 + (\lambda')^2 \pm [4(\lambda')^2 + \alpha^2]^{1/2}. \tag{92}$$

Under the condition of

$$\sqrt{1-\alpha} < \lambda' < \sqrt{1+\alpha}, \tag{93}$$

m is imaginary. If λ' is converging to 1, m becomes imaginary, and the corresponding waves are nonpropagating to be scattered.

If the $\vec{u}$ vector is right-handed in space, and

$$\vec{u} = (1/\sqrt{2})\begin{pmatrix} 1 \\ -i \end{pmatrix}, \tag{94}$$

the sacattered light is described as:

$$\vec{E} = \text{Re}\{\vec{u}E_0\exp[-i\omega\{t - (z/c)\}]\} \tag{95}$$

$$= \text{Re}\{\vec{u}E_0[\cos\{\omega(t - (z/c))\} - i\sin\{\omega(t - (z/c))\}\}. \tag{96}$$

Equation (94) is inserted into Eq. (96) yielding:

$$E_x = (E_0/\sqrt{2})\cos[\omega(t - (z/c))] \tag{97}$$

$$E_y = -(E_0/\sqrt{2})\sin[\omega(t - (z/c))]. \tag{98}$$

Equations (97) and (98) mean that the electric vector of the scattered light rotates clockwise at the origin ($z = 0$) viewing from the origin to the positive side of the z axis. This is right circularly polarized light. If $\varepsilon_0^{1/2} \cong n$ (n: refractive index of the liquid crystal), the wavelength of the scattered light, having a maximum intensity

$$\lambda_{max} = nP, \tag{99}$$

because the incident light is scattered near $\lambda' = 1$ ($\lambda' = \lambda/\varepsilon_0^{1/2} P$).

When linearly polarized light passes through a cholesteric phase having helical directors wound in a right-handed helix, RCP is scattered selectively, and the sacttering results in $\varepsilon_L \ll \varepsilon_R$, i.e., $\Delta\varepsilon < 0$. Thus, the pitch-band CD with minus sign corresponds to a right-handed helical sense in the mesophase.

7.2 The Relationships Between the ICD Signs and the Sign of the Pitch-Band CD

Saeva et al. [62] first found that ICD signs are dependent on the position of λ_{max} of the cholesteric pitch band relative to the wavelengths of the absorption bands. For the right-handed cholesteric mesophases composed of cholesteryl chloride-cholesteryl nonanoate mixtures, the ICD sign for a transition moment with a preferred orientation parallel to the long-axis of the liquid crystal molecules is plus for a longer wavelenth transition than λ_{max}, while it is minus for a shorter wavelength transition. This situation is true in the case where the long-axis of the added dye molecules is aligned along the helical directors in the cholesteric phase. For example, Fig. 70 shows the ICD of pyrene-2-carboxylic acid methyl ester intercalated into the cholesteric mesophase, which exhibits a pitch-band CD with *plus* sign near 900 nm. The plus sign means that the mesophase is left-handed. In wavelength regions shorter than 900 nm, long-axis polarized bands show the ICD bands with *plus* sign, while short-axis ones show *minus* [249].

The above-mentioned discussions have dealt with the parallel incidence to the helical axis of the cholesteric phase, not oblique incidence, in which higher-order reflections are allowed. Experimentally, most samples are aligned incompletely. Then, the CD phenomena in oblique incidence for well- or randomly-orientated quasicholesteric phases must be treated theoretically. In such cases, the linear dichroism effect on the observed CD spectral profiles may be serious and leads to difficulties in theoretical treatments. Sindo and Ohmi concluded that "liquid crystal-induced CD" cannot be related to true optical activity without careful considerations of the experiments [250]. They consider that both circularly and linearly polarized contributions are present for the helically arranged thin layers composed of optically active molecules. Typical examples of this case are CD spectra observed in a jelly of chiral 12-hydroxyoctadecanoic acid [251] and in a liquid crystal of N-acylamino acids and organic solvents [252]. The latter CD has been assumed to arise from a type of "Christiansen effect". The N-acylamino acid-solvent systems exhibit irridescent color, resulting from the difference of the optical dispersion between the solvent and the suspended liquid crystalline phase. This system can be grouped into an intermediate group between chromatic suspensions (solid-liquid systems) [254] and chromatic emulsions (liquid-liquid systems) [255]. Chromatic suspension of liquid crystals similar to the N-acylamino acid-solvent systems could be obtained when dialuminium stearate was suspended in benzene or the solvent mixture of methyl iodide and chloroform.

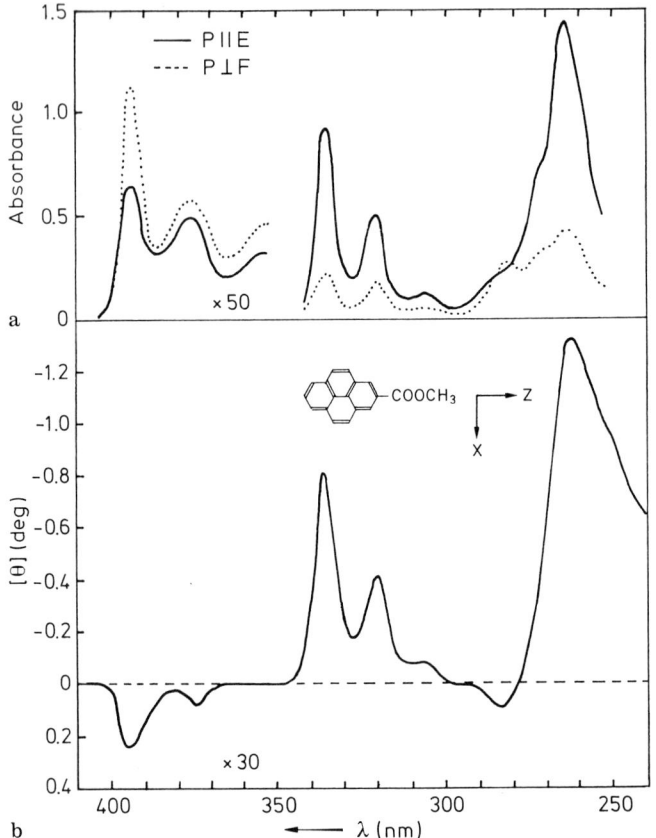

Fig. 70a and b. Absorption polarization spectra (**a**) and the ICD spectrum (**b**) of pyrene-2-carboxylic acid methyl ester [249]
For CD observation, pyrene-2-carboxylic acid methyl ester was dissolved into a cholesteric mixture of 55.5 mole percent of cholesteryl nonanoate and 44.5 mole percent of cholesteryl chloride, which exhibited a pitch band CD with plus sign

The dialuminium stearate-benzene system exhibited no optical activity because of absence of any kind of optical activity in the components. The selective transmission of the N-acylamino acid-solvent systems was also confirmed by measuring the dispersion of refractive indices in the systems. The wavelengths of the selective transmission of the light travelling through the N-acyl amino acid-solvent systems shift depending on the temperature. The observed CD maxima variation was not consistent with the maximum transmission variation. The origin of the CD observed for the N-acyl-amino acid-solvent systems may remain an issue for further discussion. However, the appearance of the induced CD bands due to achiral molecules intercalated into these systems and formation of a spherulite-like phase having an optically negative sign led to the conclusion that this suspended liquid crystalline phase has a cholesteric helical structure.

Recently, we also found that several types of chiral smectic C phases composed of achiral and chiral aromatic esters exhibit a large shift of their pitch band CD, depending on the temperature ranging from ambient to 50 °C [256]. The pitches estimated from the pitch bands are in agreement with those observed by other optical measurements. These CD phenomena require further investigating by using theoretical and experimental aids such as NMR [257].

7.3 Enhanced Circular Dichroism of Aggregates of Chiral Amphiphiles

It has been reported that a variety of single-chain amphiphiles spontaneously produce stable, membrane-forming aggregates when dispersed in water [258-260]. Dialkyl-amphiphile L-III (L or D means L- or D-configuration of amino acid unit in compound III, respectively), which was prepared by condensation of didodecyl L-glutamate and p-(4-bromobutoxy)benzoic acid and the subsequent quarterization with trimethylamine, produces bilayer vesicles in water as probed by electron microscopy [261].

$$CH_3-(CH_2)_{11}-OC-C-NC-\bigcirc\bigcirc-O-(CH_2)_4-N^{\oplus}-(CH_3)_3$$
$$CH_3-(CH_2)_{11}-OC-(CH_2)_2 \qquad Br^{\ominus}$$

III

Figure 71 shows the temperature dependence of the CD spectrum of L-III in a dilute aqueous solution (1.0×10^{-4} M). At temperatures above 31–32 °C, corresponding to the phase-transition temperature (T_c), the CD spectrum possesses an extremum at 245 nm with $[\theta]_{245} = +6000$. At temperatures below T_c, extremely large CD extrema appeared at 220 and 260 nm: $[\theta]_{220} = +360000$ and $[\theta]_{260} = -400000$ when the temperature was lowered to 15 °C. This temperature dependence of the CD was reversible, and similar spectral changes were observed for D-III in the mirror image. The CD enhancement is highly dependent on the amphiphile concentration. This concentration dependence is assumed to be related to the critical micelle concentration of III. The enhanced CD was destroyed by the addition of other surfactants such as cetyltrimethylammonium bromide at temperatures below 30 °C. These data reveal that the enormously large CD is derived from the molecular fixation of chiral surfactants in the rigid bilayer assembly. This situation may be quite similar to that in the light-harvesting bacteriochlorophyll-protein complexes embedded in the membrane assembly. Kunitake et al. also found that large CD is induced when methyl orange is bound to the similar bilayer membrane assembly of chiral dialkyl amphiphiles and the ICD magnitudes are changed drastically by their gel-to-liquid crystal transition and chemical structures [263]. Similar CD enhancements also have been found for single-chain amphiphiles containing more stiff groups such as the biphenyl or azobenzene group [264-266]. Double-chain ammonium amphiphiles without benzene group can produce typical bilayer vesicles, when dispersed in water by sonication. Upon aging at 15–20 °C for one day, twisted filaments appeared, and then one filament changes completely into a helix after incubating for several hours [266]. This phenomenon is quite similar to that of lecithines forming a helical myelin shape by addition of Ca^{2+} ions [267]. A similar helix formation also has been reported for 12-

7.4 Future Trends and Scope on Liquid Crystal-Induced Circular Dichroism

Nordén et al. have examined the CD spectral changes of dyes oriented in a liquid crystal matrix, on variing the order parameter ϱ, which is estimated by NMR observation [262]. Kuball et al. presented a theoretical description of the optical activity of oriented molecules, and they found that the CD of the transition A of a given oriented molecule $\Delta\varepsilon^A(\tilde{v})$ is described by:

$$\Delta\varepsilon^A(\tilde{v}) = (1 - p)\Delta\varepsilon_{iso}(\tilde{v}) + p\Delta\varepsilon_{33}^A(\tilde{v}), \tag{100}$$

where $\Delta\varepsilon_{iso}^A(\tilde{v}) = (\Delta\varepsilon_{11}^A(\tilde{v}) + \Delta\varepsilon_{22}^A(\tilde{v}) + \Delta\varepsilon_{33}^A(\tilde{v}))/3$ [273-276].

This equation gives an estimation of the direction of the transition moment. Again, it should be noted that the CD of an anisotropic system can be falsified by interference with the effects of linear birefringence and linear dichroism. One should have in mind that the linear birefringent or linear dichroism contributions are three orders of magnitude larger than that of the intrinsic CD.

The circular dichroic contribution from light scattering makes it difficult to estimate the CD of particulate samples such as membrane suspensions, nucleosomes, erythrocyte ghosts, or viruses. Differential light scattering is the difference in the extent of scattering of LCP and RCP light by an optically active sample [277]. The differential light scattering is diagnosed with a dependence of the CD signals, in their spectral profile and magnitudes, on the detector acceptance angle and the detector-sample distance, and the CD signals outside the absorption range of the chromophore. The angular dependence, distance dependence, and magnitude of the differential scattering intensities are functions of the particle size and the relative orientation and distance between the scattering units within each particle. A complete correction for differential scattering is practically possible by using fluorescence-detected CD, which permits measurements with an effective acceptance angle over 4π steradians [277].

On the other hand, absorption flattering occurs when chromophores are closely packed. The probability for protein molecules to encounter photons, and the absorption, is reduced in this case. The extent of flattering is a function of the size of the particles and the concentration of chromophores within each particles. Mao and Wallace [278] found that the large size of membrane particles and the high local concentration of proteins give rise to differential scattering and absorption flattering effects which result in significant distortions of the CD spectra of membrane proteins. To minimize the scattering and flattering effects several methods of fragmentation, including sonication, solubilization, and incorporation into small unilamellar vesicles, were examined and were compared in terms of the CD spectrum for each method. While sonication decreased differential scattering, it had little effect on the total distortion. Solubilization in octyl glucoside tended to decrease both differential scattering and flattering, but induced some conformational change in the protein. When the protein was incorporated into small unilamellar vesicles, the observed CD was nearly identical with the calculated one.

The needs to observe the CD of samples having ordered structures such as mitochondria and viruses will increase in the future. Much attention should be paid to CD observations of samples in which optically active molecules are aligned to each other and are concentrated in the ordered systems.

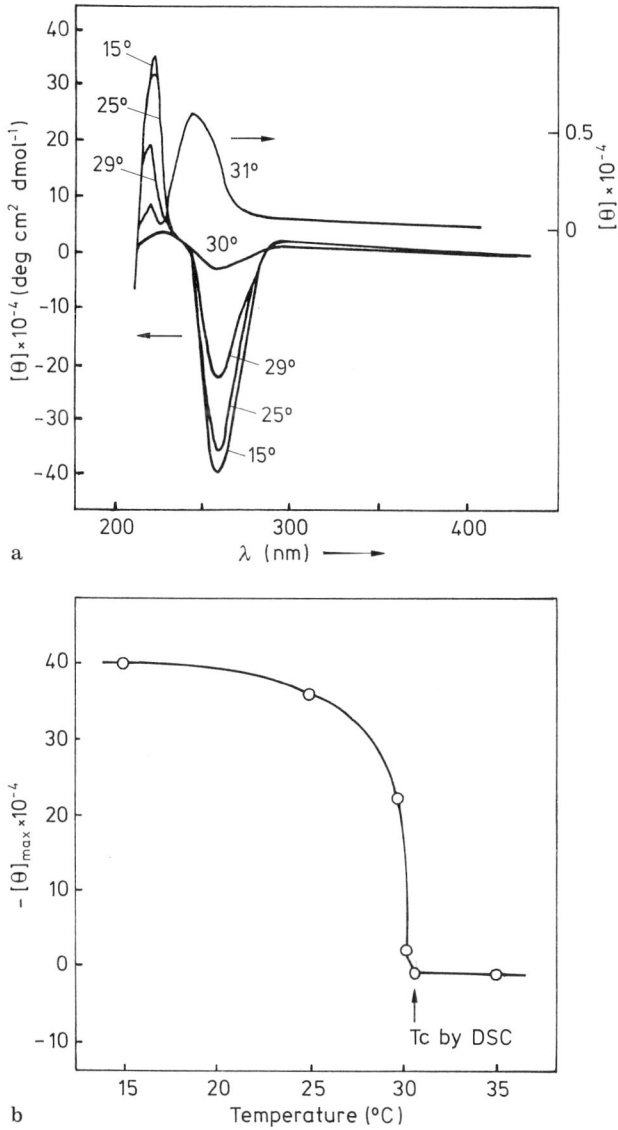

Fig. 71 a and b. CD spectral change of chiral bilayers in water [261] [L-III] = 1.0×10^{-4} M.
a. Spectral change of L-III with temperature variation.
b. Temperature dependence of $[\theta]_{max}$ for aqueous bilayers of L-III

hydroxystearic acid in both the gel state and in a collapsed monolayer [268-272]. The CD behavior is highly interesting, and further investigations along these lines are expected. Lastly, it should be noted that the observed CD enhancement may actually have arisen from the linear dichroism contribution, which has been predicted by several authors [250, 262].

In these cases, a variation of the distance between the sample and the detector, in which the variation of the detector acceptance angle is also examined, provides an easier diagnosis on the distortion of the CD resulting from the linear birefringence, linear dichroism, and light-scattering effects.

8 Experimental Considerations and Apparatus for Measuring Circular Dichroism

8.1 Experimental Considerations and Precautions

a) Samples

The samples for CD and MCD measurements must be *isotropic*. Otherwise, the sample having birefringence or dichroism in itself can be tested by rotating the sample around an axis parallel to the incident circularly polarized light. If there is some birefringence or dichroism in it, the observed CD or MCD signals will be changed or sometimes inverted in sign. A typical sample volume for a cell with a 10 mm light path length ranges from 0.6 to 3.5 ml. For small-volume sample solutions, a specially designed cell in which the sample is filled up to the light path space of 4 mm in diameter and 10 mm length, can be used. Sometimes, a well-calibrated spacer is inserted in the cell. The concentration of sample in a solution must be adjusted in the range of absorbance from 0.5 to 1.3. For a sample solution with absorbance lower than 0.5, CD signals are often quite low, and occasionally CD signals are superimposed with noise levels. On observing ICD of an achiral molecule mixed with a chiral molecule in solution, the signal-to-noise-problem becomes serious. This problem can be circumvented by many repetitive scannings followed by signal accumulation. This function is often combined with curve smoothing by using a computer.

Most CD or MCD measurements are performed on dilute solutions. But, measurements of CD or MCD spectra in the near infrared region requires very high concentrations of the samples. For example, we have used solutions of myoglobin derivatives with molar concentrations above 0.3 mM for their MCD in the near infrared region. These samples of biopolymers at high concentrations can be obtained by using a centrifuge (3000 rpm) together with an Amicon CF-25 Centriflo^R (Amicon Corp., Lexington, Mass.) membrane cone. The membrane cones also are suitable for removing dust in a sample solution.

Crystalline or solid samples (ca. 10 mg) are ground with 5 drops of nujol in an agate mortar to form a paste for subsequent CD or MCD measurement. Perfluorobutadiene can also be used instead of nujol. Otherwise, the KBr pellet technique, for which KBr and sample should be completely dried *in vacuo* and ground into a fine powder, can be used conveniently. On using the KBr technique, one can often observe a different CD spectrum from that of the nujol-mull method.

b) Cells

Many types of cells having a variety of path lenths are commercially available. In our laboratory, various types of well-designed cells are used for CD measurements at lower temperatures. For example, a cell having a long guide tube and a reaction vessel capable of evacuation, is very useful for CD and MCD measurements at the temperature of liquid nitrogen. For MCD and CD measurements at the temperature of liquid helium, an Oxford superconduction magnet (SCM 4 type) is quite useful, because the sample temperature can be kept constant and adjusted from 4 K to ambient by boiling off helium gas and by its internal heater.

An alternative type of cell, which consists of two parts of optically flat windows, is suitable for CD and MCD measurements of small-volume samples. One of the window affords a trough for filling in the sample. Otherwise, a well-calibrated spacer is inserted to a conventional cell for adjusting its path length. The light path length is calibrated by using the absorbance of an appropriately diluted solution of benzene or toluene.

c) Calibration of CD and MCD Magnitudes

The magnitude of the CD has commonly been calibrated at 290 nm using (+)-camphor-10-sulfonic acid as a standard. Because of its hygroscopic nature, the water content has led to some confusion. Tsujimura et al. have recommended to use a molar ellipticity of $+7.78 \times 10^3$ for the compound in an aqueous solution [280]. The concentration can be calibrated by the ORD of the same sample, which gave a molar rotation of $+4.28 \times 10^3$ at 305 nm and of -5.44×10^3 at 270 nm.

In order to eliminate difficulty in the preparation of (+)-camphor-10-sulfonic acid, a proposal has been made to use D-pantolactone and (+)-tris(thylenediamine)-cobalt(III) iodide monohydrate as calibration standards at 220 and 490 nm, respectively. For the former having $[\alpha]_D^{23} = -51.0°$ ($c = 2$ in water), the molar ellipticity at 220 nm is -13.1×10^3 in methanol (-17.3×10^3 in water) and the [θ] value at 490 nm for the latter in water is $+6.24 \times 10^3$ ($[\alpha]_D^{23} = +89°$ in water). Androsterone, having a molar ellipticity at 304 nm of 11.2×10^3 degree cm^2 decimol^{-1} in dioxane, is also used as a standard.

The reported CD magnitudes sometimes have given large deviations after careful calibration using (+)-camphor-10-sulfonic acid. A test sample method revealed that the deviation of molar ellipticity became considerably large with increasing the wavelength from 400 to 800 nm. For the extrema of nickel(II)tartrate at 718 and 777 nm in an aqueous solution, the deviations of the [θ] values reached to ± 35%, even if all instruments were calibrated at 290 nm using (+)-camphor-10-sulfonic acid. Several instruments showed large shifts of the sample's CD extrema at 718 and 777 nm; for example, one instrument gave a red shift of 18 nm and 24 nm for the extrema, respectively. The wavelengths of the CD extrema of the nickel(II) tartrate solution should be calibrated with a standard filter composed of neodymium glass, referring the absorption to the HT-voltage and/or the DC current of the dichrometers. Furthermore, the absorption maxima of the neodymium glass filter are calibrated at the wavelengths of the line spectrum of the D_2-lamp (at 656.10 nm and 486.00 nm). Corrected wavelengths of the CD extrema of the nickel(II) tartrate solution are 371, 399, 428, 470, 718, and 777 nm. In conclusion, it is recommended to use the [θ] values of the nickel(II) tartrate solution, -101.2 at 718 nm and -110.7 at 777 nm (in degree dm^3 mol^{-1} cm^{-1}), as a CD standard in the longer wavelength region [281].

For MCD measurements, the instrument is calibrated using the natural CD of D-10-camphorsulfonic acid as a standard. The magnetic field is determined with freshly prepared hexacyanoferrate ion, $\Delta\varepsilon_{422}/H = 3.0 \,(\text{cm M T})^{-1}$, where 1T = 10000 G. When the direction of the magnetic field is parallel to the Poynting vector, it is positive. As far as this convention is held, the Verdet constant (see Appendix) of water is negative. To separate the natural CD from each MCD spectrum, the spectra are recorded twice once with the magnetic field parallel and then without any magnetic field.

d) Temperature Variation

For the CD and MCD measurements at temperatures below 0 °C, a cell of 2 mm path with a copper-constantan thermocouple is subjected to a stream of cold nitrogen gas or liquid nitrogen in a Dewar vessel enclosed by a thermostatted heater. Glassy matrices are used as solvents for CD measurements at -196 °C [282]. In a similar way, polymer film such as polyvinylalcohol or polyethylene film including the sample can be satisfactorily used. On using these matrices, one should check whether a contribution of linear dichroism or birefringence to the observed CD is present or not. Especially, in measurements of biological samples, glassy solvents such as glycerol (up to 70% in volume)-H_2O or potassium glycerophosphate(KGP)-glycerol-H_2O are suitable.

The wide variation of temperatures from ambient to 4.2 K can be conducted easily by using a superconducting magnet, Oxford SCM-4, in which the sample temperature is kept constant by evaporated He gas and a heater system.

For CD and MCD observations at temperatures above ambient, many types of cells equipped with jackets through which thermostatting fluid can be circulated from an external constant-temperature bath are useful. Otherwise, a thermostatting arrangement involving a sample holder through which fluid from a constant-temperature bath can be circulated is used. In the latter case, it can take up to 30 min for the system to equilibrate thermally. Thermal fluctuation often causes a drift of the base-line.

8.2 Apparatuses

a) Monochromators

Almost all commercial spectropolarimeters use dual quartz monochromators. A double monochromator is essential to keep stray light at an absolute minimum and the wavelength resolution in the highest level. Two standard models of Jasco J-500 spectropolarimeters cover the wide wavelength ranges of 180–700 nm (J-500A) and 180–1000 nm (J-500C). In J-500A, a few of the single quartz crystals with different axes are utilized for obtaining linearly polarized light. In the latter (J-500C), a Rochon prism is added to the optical system for assuring linearly polarized light in the longer wavelength range. Accordingly, the former provides a better signal-to-noise ratio in the shorter wavelenth region than the latter does.

The spectropolarimeters for the CD and MCD observations in the vaccuum ultraviolet region use prisms and a Rochon prism made of CaF_2 or MgF_2 [234].

The monochromator for the CD in the infrared region is constructed with a grating, because a grating is more suitable for the optical element in the infrared region. For example, a Jasco J-200 uses a single monochromator incorporating a 300 lines/mm grating blazed at 1000 nm [283]. Nafie et al. used an infrared CD spectropolarimeter equipped with a 3/4-m monochromator (Spex Model 1702) having a 300 lines/mm grating blazed at 3000 nm [284]. The infrared CD observation can also be achieved with Fourier transform infrared spectrometer [285] instead of a dispersive, grating spectrometer. The optical energy in the infrared region is not sufficient to obtain a good signal-to-noise ratio, and then one is compelled to use a single monochromator incorporating a grating or to use a Fourier transform spectrometer.

b) Polarizers and Stress Modulators

The performance of the latest spectropolarimeters is enhanced significantly by the use of stress modulators (photoelastic modulators), replaced electrooptical modulators (Pockells' cells) [286]. For the vacuum ultraviolet CD, a CaF_2 modulator is utilized, because it is isotropic and transparent to about 125 nm. The commercially available spectropolarimeters such as the Jasco J-500 are equipped with modulators made of quartz. A Jasco J-200 spectropolarimeter, scanning in the range from 900 to 2400 nm, is incorporated with a CaF_2 modulator. Since the transparency of CaF_2 is limited to 7 μ and is brittle, a single crystal of Ge is utilized as a stress modulator for the infrared CD ranging from 2 to 10 μ [287].

In a similar way, Rochon prisms are made of CaF_2 or MgF_2 for the vacuum ultraviolet, and quartz for the ultraviolet to visible region. For infrared CD, a wire-grid is used as a polarizer.

c) Detectors

The photomultipliers equipped in the Jasco J-500A and Jasco 500C are Hamatsu R-376 and R-316, respectively. The latter is of S-1 type. An InSb photovoltaic cell (Judson) is used as a detector for the wavelength region from 1000 to 2400 nm [283]. The InSb detector is cooled with liquid nitrogen. An extension of the CD measurements to 11 μ can be conducted by using HgCdTe detectors cooled at liquid He temperature [287].

d) Improvement of the Signal-to-Noise Ratio

The signal-to-noise ratio of the CD measurements depends on the amount of light reaching the detector. This is adjusted by changing the mechanical slit widths with concomitant changes in resolution, which is approximately proportional to the inverse cube of the wavelength. Then, most commercial spectropolarimeters are programmed on slit control, providing the reproducible band shape and CD magnitude. One should scan the spectrum a second time after reducing the mechanical slit width by a factor of two. The band shape and CD magnitudes must be independent of the slit width.

Deterioration of the Xe lamp and mirrors in the monochromator causes the signal-to-noise ratio to go down. Especially, it is serious for measurements at shorter wavelengths than 250 nm. The decrease of the reflectance on the mirrors is prevented easily by purging nitrogen gas into the monochromator in operating the spectropolarimeter. The mutual configuration, including the height, should be adjusted to ensure the best condition.

The high-frequency noise can be diminished either by slowing the time constant of the amplification circuitry or by averaging the signals on the computer. The low-frequency noise is caused by thermal fluctuation in the instrument. If one observes the CD spectrum with extremely small CD magnitudes, a two or three hour warmup is needed to stabilize the base line. Deterioration of the Xe lamp or mirrors often gives rise to drifting of the base line.

The light sources used for CD measurements must be intense and should possess good short-term stability. Synchrotron radiation promises to be an excellent source of radiation for CD measurements in the wavelength range from vacuum ultraviolet to infrared [247,248].

9 Concluding Remarks and Future Trends on Induced Circular Dichroism

Most biopolymers wind in the right-handed sense. The origin of this handness in biopolymers lies in the fact that the chiral monomeric units, nucleotide or α amino acid residues, are all of one-type: D-enantiomers of sugars for nucleotides and L-enantiomers of amino acids for proteins. This handness may be needed for hereditary continuity and to discriminate the enantiomers of all metabolites, which is essential for physiological reactions. Nucleotide sequences in DNA and RNA or amino acid sequences in proteins are designed in nature to assume the most favorable structure in space for their hereditary and physiological reactions. Recent gene manipulation techniques providing point-directed mutation may be able to gain access to clarify the design principles of nucleic acids and proteins in nature.

To reach our goal, we need more detailed and precise information on changes in the
1) conformation of nucleic acids and proteins,
2) rigidity of side chains in proteins,
3) binding mode between substrates and proteins or nucleic acid base moieties,
4) other chemical events such as tautomerism and proton dissociation,
5) chiral environments around metal ions or metal complex units.

Most CD measurements have been examined conventionally for the samples under equilibrium conditions. Equilibrium methods can only detect the species that are stably populated in the transition region [289]. If a protein possesses some intermediate states with higher stabilizing energies, the intermediate species detected kinetically are practically non-detectable by equilibrium methods. It has been well known that most small globular proteins exhibit a two-state unfolding. Thus, kinetic CD observations have become indispensable for detecting unequilibrated intermediates. For chemical reactions between substrates and enzymes the CD changes should be detected kinetically.

The time domain in observation of emission or absorption has been extended to the range of 10^{-12} or 10^{-9} s. But, CD observations are limited to the time domain longer than ms.

Stronger light sources must improve the signal-to-noise ratio for CD measurements, even if the measurements will be carried out on commercially available spectropolarimeters. Synchrotron orbital radiation (SOR) provides very strong light sources ranging from the vacuum ultraviolet to infrared region, because the SOR is emitted as a type of linearly polarized light even if it is in a pulse mode being not suitable for CD measurements in the vacuum ultraviolet region, but they are not conventional for those in the ultraviolet and visible regions. We should elaborate a new type of spectropolarimeter having higher precision, a higher signal-to-noise ratio, and a shorter time domain within μs or ns. Fourier transform (FT) data-processing has been utilized for CD measurements in the infrared region. The FT is very effective for cumulative integrations of one spectral datum *in situ*, resulting in much improvement of the signal-to-noise ratio. Further, the FT mode acquisition may be suitable for detecting the signals from an optical device using a light source in a pulse mode.

The application of a SOR light source and FT mode acquisition to CD measurements will lead to further developments in ICD research relating to biologically important phenomena.

Finally, it should be noted that other spectral analyses such as Raman, nuclear magnetic resonance, and fluorescence spectroscopies, which are correlated to more local chemical entities or individual atoms in the molecular systems, complement the CD data. Although other spectral approaches may provide more effective tools for analyzing a local structure or individual atoms in the molecular systems than the CD does, the CD approach is indispensable for the estimation of change or population in chiral structures on the *average*.

10 Appendix: Magnetic Circular Dichroism Techniques Coupled with the Circular Dichroism Technique

The "*Faraday effect*", discovered by Faraday in 1845, refers to a phenomenon of optical rotation induced by a magnetic field set parallel to the observing light path. The magnetic optical activity is induced in all matter. *Magnetic rotation* (MOR; the Faraday effect) and *magnetic circular dichroism* (MCD) arise, respectively, from the difference in refractive indices and absorption coefficients of LCP and RCP due to the applied magnetic field [see Eqs. (15) and (30) for natural optical activity]. MOR occurs in both transparent and absorbing spectral regions, whereas MCD is in an absorptive wavelength region. MCD is the difference between left and right circularly polarized Zeeman spectra, and provides valuable information on the electronic states of a given molecule. In 1966, Buckingham and Stephens [290] reported a review on the Faraday effect and gave a detailed theoretical treatment of the dispersion of the Faraday effect through absorption bands. In this review, they confirmed that the dispersion of the Faraday effect through an absorption band can provide information not available from the absorption and should become an important tool in spectral analysis and the investigation of molecular structure. After that, reviews which further emphasize the utility of the Faraday effect or MCD in the fields of chemistry, physics, and biophysics, have been published by Schatz and McCaffery [291], Caldwell, Thorne, and Eyring [292], and Stephens [293]. Today, MOR and MCD have been widely used not only for interpretation of ambiguous and complicated absorption spectra, but also for the investigations of the magnetic properties in the ground and excited states of ions and molecules.

The experimental data of MCD are analyzed in terms of the Faraday A, B, and C terms. The Faraday A term changes the sign at the absorption maximum, while the Faraday B and C terms have the same dispersion form as the corresponding absorption. The Faraday C term is inversely proportional to the absolute temperature, whereas the Faraday A and B terms are temperature independent. Accordingly, the Faraday C term is extracted from the MCD spectrum when the observing temperature is varied. When either the ground or the excited state of a molecule or ion is degenerate, the Faraday A term should be observed. The Faraday A term presents information about the magnetic moment in the degenerate ground or excited state. The degeneracy of the ground state causes the Faraday C term, which permits us to determine the magnetic moment in the ground state. In the absence of an n-fold axis ($n \geq 3$), one can expect only the Faraday B term, which arises from the mixing of the wave-functions of the ground and excited states with those of all other electronic states by the external magnetic field. With $[\theta]_M$ in units of degree deciliter decimeter^{-1} mol^{-1} Gauss^{-1} (degree 1 m^{-1} mol^{-1} G^{-1} or degree 1 m^{-1} mol^{-1} 10^{-3} T; T, Tesla = 10^4 G), the MCD spectrum $[\theta]_M$ as a function of the wave-number ν can be expressed for *isotropic* samples as:

$$[\theta]_M = -21.35 \{f_1 A + f_2(B + C/kT)\}, \qquad (A\text{-}1)$$

where $f_1(\nu)$ and $f_2(\nu)$ are the lineshape functions for the transition, k the Boltzmann's constant, and T the absolute temperature. The Faraday A, B, and C/kT are in unit of square Debye × β, square Debye × β/cm^{-1}, and square Debye × β/cm^{-1}, respectively. The units are Debye for the electric dipole moment, and the Bohr magneton β for the magnetic dipole moment and wave-number $\tilde{\nu}$, in cm^{-1}, as a measure of energy. The Faraday A, B, and C terms are described as:

$$A = (1/2 d_a) \sum_{a \to j} [\langle j|\vec{m}|j\rangle - \langle a|\vec{m}|a\rangle] \cdot Im \{\langle a|\vec{\mu}|j\rangle \times \langle j|\vec{\mu}|a\rangle\} \qquad (A\text{-}2)$$

$$B = (1/2 d_a) \sum_{a \to j} Im \left\{ \sum_{k \neq a} [\langle k|\vec{m}|a\rangle/(E_k - E_a)] \cdot [\langle a|\vec{\mu}|j\rangle \times \langle j|\vec{\mu}|k\rangle] \right.$$

$$\left. + \sum_{k \neq j} [\langle j|\vec{m}|k\rangle/(E_k - E_j)] \cdot \{[\langle a|\vec{\mu}|j\rangle \times \langle k|\vec{\mu}|a\rangle]\} \right. \qquad (A\text{-}3)$$

$$C = (1/2 d_a) \sum_{a \to j} \langle a|\vec{m}|a\rangle \cdot Im \{\langle a|\vec{\mu}|j\rangle \times \langle j|\vec{\mu}|a\rangle\}. \qquad (A\text{-}4)$$

Here, d_a is the degeneracy on the ground state a, and the expressions (A-2)–(A-4) are given for the transition $a \to j$. E_a, E_j, and E_k are energies in cm^{-1} of the states a, j, and k, respectively. It is very important to note that a positive MCD peak corresponds to a negative B term and a negative MCD peak to a positive B term. The shape $f_2(\tilde{\nu})$ is that of an absorption peak, and the shape $f_1(\tilde{\nu})$ is that of a derivative of an absorption peak (a positive extremum followed by a negative one at a higher energy). Typical shapes are shown in Fig. A1. It is essential to obtain the quantitative values for the Faraday A, B, and C terms from the experimental data. The method of moments[291,293,294] is a powerful technique for extracting the Faraday parameters. In this method, the n-th moments are computed by numerically integrating the MCD

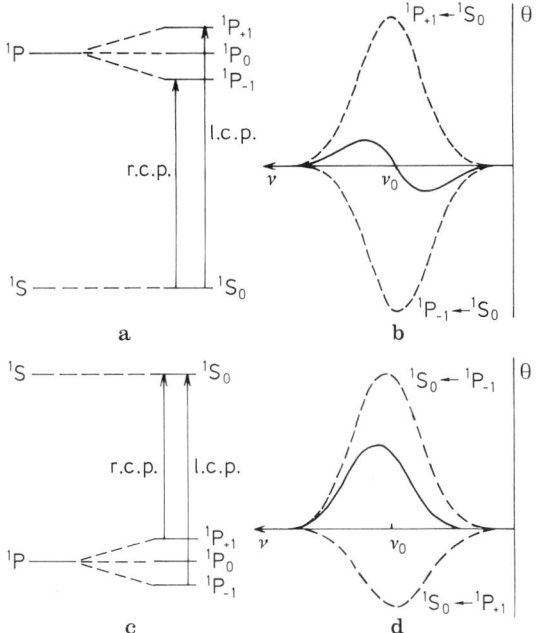

Fig. A1 a–d.
a. Zeeman splitting of the ^1P state in an external magnetic field and selection rules for circularly polarized light.
b. The effects of the Zeeman splitting in the ^1P state on MCD of $^1P \leftarrow {}^1S$ transition.
c. Zeeman splitting of the ^1P state in an external magnetic field and selection rules for circularly polarized light.
d. The effects of the differences in population of the $^1P_\pm$ states on MCD of $^1S \leftarrow {}^1P$ transition

and absorption spectra in accordance with the following equations:

$$\langle \varepsilon \rangle_n = \int_{\text{band}} (\varepsilon/\tilde{v}) \, \tilde{v}^n \, d\tilde{v} \tag{A-5}$$

$$\langle \theta_M \rangle_n = \int_{\text{band}} ([\theta]_M/\tilde{v}) \, \tilde{v}^n \, d\tilde{v}, \tag{A-6}$$

where $[\theta]_M$ and ε denote the molar ellipticity in degree deciliter decimeter^{-1} mol^{-1} Gauss^{-1} and the molar extinction coefficient in liter mol^{-1} cm^{-1}, respectively, and v is the wave-number. From Eqs. (A-5) and (A-6), one obtains:

$$\langle \theta_M \rangle_0 = -33.53(B + C/kT) \tag{A-7}$$

$$\langle \theta_M \rangle_1 = 33.53[A - (B + C/kT)\tilde{v}] \tag{A-8}$$

$$\tilde{v} = \langle \varepsilon \rangle_1 / \langle \varepsilon \rangle_0, \tag{A-9}$$

where A and $(B + C/kT)$ are expressed in square Debye × β units and in square Debye × β per cm^{-1} units, respectively. Thus, the Faraday parameters can be obtained by simple numerical integrations of the experimental data.

If both of B and C are present, a measurement of the temperature dependence of the MCD spectrum leads us to separate them. Two partially overlapping B terms of opposite signs cannot be readily distinguished from an A term, and then a couple of the B terms is called an apparent or a pseudo-A term [295].

The magnetic moment $\vec{m}$ of a degenerate excited state is expressed as [295]:

$$\vec{m} = -A/D, \tag{A-10}$$

where D is the dipole strength, which is estimated from the absorption.

For the $^1E_{1u} \leftarrow {}^1A_{1g}$ or $^1E' \leftarrow {}^1A'_1$ transitions of various hexa- and tri-substituted benzene derivatives, the A/D and B/D values were obtained by using the method of moments [291, 293, 294] from the observed MCD and absorption spectra [296-298]. These values are shown in Table A1. The electronic transition energies were calculated within the framework of a PPP approximation, and the calculated values are listed in Table A2. For the benzene derivatives with electron-donating substituents such as halogens, hydroxy-, and methoxy-groups, the calculated magnetic moments are in good agreement with experimental values both in sign and in magnitudes. The magnitudes of the calculated magnetic moments of the benzene derivatives with electron-withdrawing groups such as cyano, nitro, and carbonyl groups are much smaller than those of benzene derivatives with electron-donating substituents. This explains well the experimental results. According to a simple molecular orbital consideration [297, 298], it is clear that the molecular orbitals of electron-withdrawing groups are preferentially mixed with the lowest vacant orbitals rather than the highest occupied orbitals of the benzene ring, leading to a decrease of the magnetic moments in the $^1E_{1u}$ and $^1E'$ states of the benzene derivatives with electron-withdrawing groups. For the monosubstituted benzene derivatives, the MCD approaches are also very effective to analyze their electronic states [297, 298]. The 1st and 2nd transitions in the lower energy side are observed in the wavelength regions of 33000–40000 cm^{-1} and 37000–50000 cm^{-1}, respectively. In the case of aniline, phenol, and anisole, which are benzene derivatives with an electron-donating group, the negative and positive MCD bands are observed for the 1st and 2nd transitions, respectively. Benzonitrile, benzaldehyde, benzoic acid, and nitrobenzene show positive and negative MCD bands for the 1st and 2nd transitions, respectively, which are typical of the benzene derivatives with electron-withdrawing group. The observed and calculated

Table A1. Extract Faraday parameters for $^1E_{1u} \leftarrow {}^1A_{1g}$, $^1E' \leftarrow {}^1A'_1$, and $^1E' \leftarrow {}^1A'$ transitions of benzene derivatives with D_{6h}, D_{3h}, and C_{3h} symmetry [296]

Compd	$A/D, \beta$	$B/D, \beta/\text{cm}^{-1}$
Hexachlorobenzene	0.182	-1.98×10^{-5}
Hexabromobenzene	0.207	-4.45×10^{-5}
1,3,5-Trichlorobenzene	0.323	-5.79×10^{-5}
1,3,5-Tribromobenzene	0.301	-7.54×10^{-5}
Phloroglucin	0.127	-1.29×10^{-5}
1,3,5-Trimethoxybenzene	0.262	2.49×10^{-5}
1,3,5-Tricyanobenzene	0.172	-2.35×10^{-5}

Table A2. Observed and calculated transition energies [296]

Compd	v_{obsd}, eV	v_{calcd}, eV	Compd	v_{obsd}, eV	v_{cald}, eV
Hexachloro-		4.21	1,3,5-Tricyano-	4.34	4.41
benzene	5.34	5.07	benzene	5.45	5.39
	5.71	5.79		5.81	5.74
Hexabromo-		4.21	1,3,5-Trinitro-	3.75[a]	
benzene	4.94	5.02	benzene	4.28	4.26
	5.43	5.68		5.06	5.08
				5.61	5.30
1,3,5-Trichloro-	4.54	4.53			
benzene	5.60	5.58	1,3,5-Benzenetri-	4.39	4.39
	6.18	6.17	carboxylic	5.35	5.28
			acid	5.85	5.62
1,3,5-Tribromo-	4.50	4.51			
benzene	5.52	5.51	1,3,5-Benzenetri-	4.26	4.39
	5.88	6.02	carbonyl	5.06	5.28
			chloride	5.47	5.64
Phloroglucin	4.63	4.42			
	5.64	5.55	Benzene	4.89[b]	4.91
	6.12	6.18		6.17[b]	6.20
				6.98[b]	7.03
1,3,5-Tri-	4.68	4.40			
methoxybenzene	5.46	5.49			
	6.04	6.10			

[a] $n \to \pi^*$ transition of nitro group;
[b] Kimura, K., Nagakura, S.: Mol. Phys. 9, 117 (1965)

B terms of the 1st and 2nd $\pi \to \pi^*$ transitions of the mono-substituted benzenes are summarized in Table A3. The major parts of the B terms of the 1st and 2nd $\pi \to \pi^*$ transitions may come from the mixing with the lowest four or five excited electronic states, because the terms with larger denominators in Eq. (A-3) are considered to be negligibly small. The contribution to the B terms comes from the mixing between the first two excited states, resulting in the opposite signs of the MCD bands, as shown in Table A4 and Fig. A2.

On benzene and some other aromatic systems, the donor or acceptor nature of a substituent is critical for the MCD sign, whereas on others such as pyridine it is not, at least for weak substituents. Further, the MCD is helpful for the analysis of a wide field of chemical events such as conformational changes coupled with substitutions, proton tautomerism, proton dissociation on aromatics, or through-space interaction in bicyclic aromatic systems [299-318].

As mentioned above, the MCD is quite useful for the
1) assigments of the observed absorptions,
2) differentiation of $\pi \to \pi^*$ transitions from others such as $n \to \pi^*$,
3) estimation of the mixing character of a given substituent's orbitals with the highest occupied or the lowest vacant orbitals of the parent aromatics,

and then the MCD diagnosis can also lead to clarify the

Table A3. Observed and calculated Faraday B values of the first (a) and second (b) $\pi \to \pi^*$ transitions of monosubstituted benzenes (10^{-5} $\beta D^2/cm^{-1}$) [297)]

(a)

	B_{obsd}	$B_{calcd}(r)^a$	$B_{calcd}(\nabla)^a$
Aniline	49.0	136.6	49.2
Phenol	18.3	66.3	24.2
Anisole	29.9	73.0	27.1
Styrene		−0.5	1.0
Benzonitrile	−18.9	−33.5	−13.7
Benzaldehyde	−37.5	−86.1	−32.3
Benzoic acid	−37.5	−47.8	−18.0
Nitrobenzene	−46.0	−166.6	−49.4

(b)

	B_{obsd}	$B_{calcd}(r)^a$	$B_{calcd}(\nabla)^a$
Aniline	−94.5	−233.3	−84.6
Phenol		−157.4	−60.8
Anisole	−27.5	−162.8	−64.6
Styrene	53.8	45.9	19.6
Benzonitrile	102.9	113.0	51.3
Benzaldehyde	118.1	173.0	69.1
Benzoic acid	130.2	141.4	56.3
Nitrobenzene	93.3	257.0	78.4

[a] $B_{calcd}(r)$ and $B_{calcd}(\nabla)$ are the calculated B values by a dipole length method and dipole velocity method, respectively

Table A4. Theoretical $B^k_{a \to j}$ values of the first and second $\pi \to \pi^*$ transitions of monosubstituted benzenes (10^{-5} $\beta D^2/cm^{-1}$) [297)]

	j/k	1	2	3	4	5
Aniline	1	0.0	60.8	0.0	−18.5	
	2	−60.8	0.0	−24.6	0.0	
Phenol	1	0.0	35.8	0.0	−14.1	
	2	−35.8	0.0	−27.3	0.0	
Anisole	1	0.0	40.5	−0.1	−16.2	
	2	−40.5	0.0	−27.2	0.0	
Styrene	1	0.0	1.3	−0.2	−0.3	0.0
	2	−1.3	0.0	15.6	2.1	0.0
Benzonitrile	1	0.0	−22.5	0.0	9.5	
	2	22.5	0.0	22.7	0.0	
Benzaldehyde	1	9.0	−43.9	−0.9	14.0	
	2	43.9	0.0	12.0	−0.3	
Benzoic acid	1	0.0	−28.3	0.6	1.4	9.4
	2	28.3	0.0	3.8	20.4	0.8
Nitrobenzene	1	0.0	−58.6	0.0	0.0	13.6
	2	58.6	0.0	5.5	1.8	0.0

[a] Calculated at the center of charge by use of ∇ operator

Induced Circular Dichroism in Biopolymer-Dye Systems

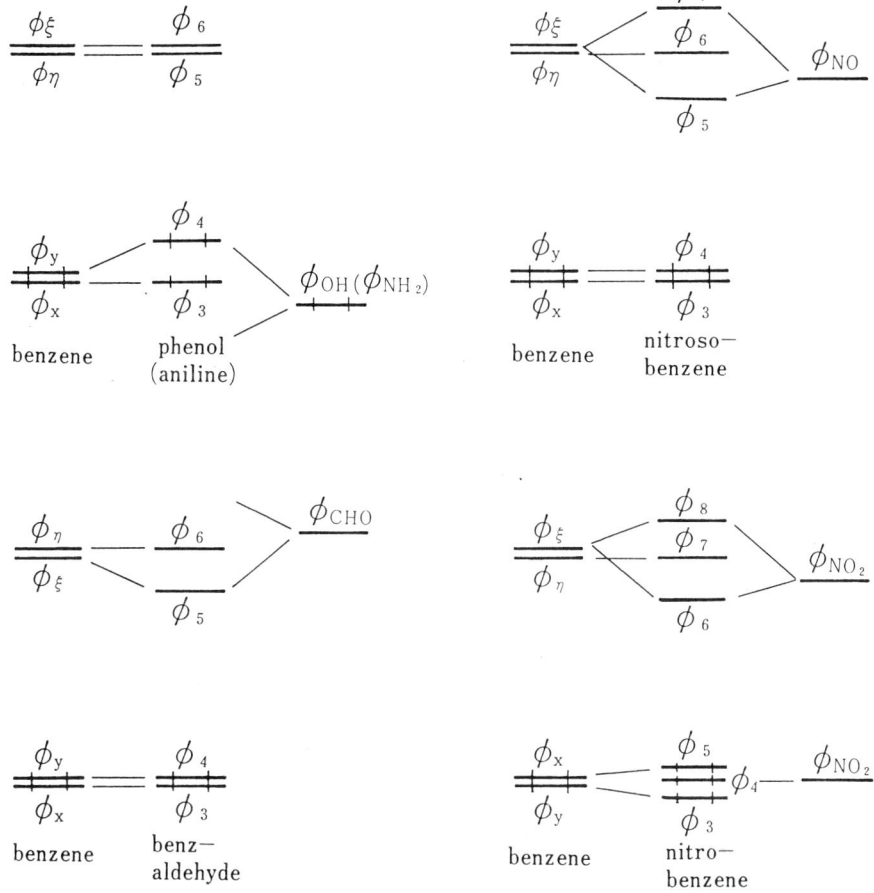

Fig. A2. Interaction of the HOMOs and LUMOs of the benzene ring with the orbitals of the substituent group [298]

4) degeneracy and effective symmetry character of a given electronic state,
5) polarization direction of observed transitions,
and
6) effective symmetry of the chromophores or aromatic moieties.

The latter three, (4)–(6), are utilized for qualitative elucidation of the observed ICD bands in signs, and the former three, (1)–(3), are successfully applicable for quantitative calculations of the observed ICD bands in both signs and magnitudes. The ICD of β-cyclodextrin complexes with benzene derivatives or azanaphthalenes [162] has been analyzed by a molecular orbital calculation, using an approximation of PPP-type, which has been compared with the theoretical spectra calculated by using the CNDO/S-CI method on the basis of the MCD spectra [165].

The MCD is independent of the chiral environment. Accordingly, it is advantageous to simultaneously observe the MCD spectrum together with the CD.

Finally, I would like to cite the following books and reviews as references for further reading on MCD.

a) Hemoproteins:
Hatano, M., Nozawa, T.: "Magnetic Circular Dichroism Approaches to Hemoprotein Analyses", in Adv. Biophys., Vol. 11, Yomosa, S., Hatano, M., eds., Japan Scientific Societies Press, Tokyo and University Park Press, Baltimore, 1978, pp. 95–152

b) Aromatic Molecules:
Michl, J.: "Magnetic Circular Dichroism of Aromatic Molecules", Tetrahedron *40*, 3845–3934 (1984)

c) Metal Ions and Metal Complexes:
Piepho, S. B., Schatz, P. N.: "Group Theory in Spectroscopy; with Applications to Magnetic Circular Dichroism", A Wiley-Interscience Publication, John Wiley & Sons, New York, 1983

d) Biological Molecules:
Sutherland, J. C., Holmquist, B.: Annu. Rev. Biophys. Bioeng. *9*, 293–326 (1980)

Acknowledgements: The author sincerely thanks Dr. Seizo Okamura of Emeritus Professor of Kyoto University and Dr. Almut Heinrich of Springer-Verlag for their continuous and genial encouragements. And the author wishes to thank the Societies and the Publishers for their permissions to reproduce the cited figures in this manuscript.

11 References

1. Mason, S. F., Norman, B. J.: Chem. Commun. 335 (1965)
2. Bosnich, B.: J. Amer. Chem. Soc. *88*, 2606 (1966)
3. Bosnich, B.: J. Amer. Chem. Soc. *89*, 6143 (1967)
4. Hayward, L. D., Totty, R. N.: Chem. Commun. 676 (1969)
5. Noack, K.: Helv. Chim. Acta *52*, 2501 (1969)
6. Axelrod, E., Barth, G., Bunnenberg, E.: Tetrahedron Lett. 5031 (1969)
7. Bolard, J.: J. Chim. Phys., Physicochim. Biol. *66*, 389 (1969)
8. Bolard, J.: J. Chim. Phys., Physicochim. Biol. *67*, 550 (1970)
9. Hayward, L. D., Totty, R. N.: Can. J. Chem. *49*, 624 (1971)
10. Hayward, L. D.: Chem. Phys. Lett. *33*, 53 (1975)
11. Schipper, P. E., Nordén, B.: Chem. Phys. Lett. *67*, 99 (1979)
12. Perrin, J. H., Hart, P. A.: J. Pharm. Sci. *59*, 431 (1970)
13. Clarke, D., Grainger, J. F.: "Polarized Light and Optical Measurements", Pergamon Press, Oxford (1971), p. 82
14. Rosenfeld, L.: Z. Physik. *52*, 161 (1928)
15. Pasteur, L.: Compt. Rend. Paris *26*, 535 (1848)
16. Werner, A.: Chem. Ber. *44*, 1887 (1911)
17. MacCullagh, J.: Phil. Mag. (3) *10*, 42, 382 (1837)
18. Kuhn, W.: Trans. Faraday Soc. *26*, 293 (1930)
19. Caldwell, D. J., Eyring, H.: "The Theory of Optical Activity", Wiley-Interscience, New York (1971)
20. Kasha, M.: Radiation Res. *20*, 55 (1963)
21. Mason, S. F., Vane, G. W.: J. Chem. Soc. *B*, 370 (1966)
22. Barnett, C. J., Drake, A. F., Mason, S. F.: Bull. Soc. Chem. Belg. *88*, 853 (1979)
23. Harmor, T. A., Robertson, J. M.: J. Chem. Soc. 194 (1962)
24. Harada, N., Uda, H.: J. Amer. Chem. Soc. *100*, 8022 (1978)

25. Rogers, D., Unal, G. G., Williams, D. J., Ley, S. V., Sim, G. A., Joshi, B. S., Ravindranath, K. R.: J. Chem. Soc. Chem. Commun. 97 (1979)
26. Harada, N., Nakanishi, K.: "Circular Dichroic Spectroscopy; Exciton Coupling in Organic and Bioorganic Stereochemistry", University Science Books, Mill Valley, California (1982)
27. Hagishita, S., Kuriyama, K., Hayashi, M., Nakano, Y., Shingu, K., Nakagawa, M.: Bull. Chem. Soc. Jpn. *44*, 496 (1971)
28. Tanaka, J., Ozeki-Minakata, K., Ogura, F., Nakagawa, M.: Nature, Phys. Sci *241*, 22 (1973)
29. Brongersma, H. H., Mul, P. M.: Chem. Phys. Lett. *19*, 217 (1973)
30. Mason, S. F.: J. Chem. Soc. Chem. Commun. 239 (1973)
31. Kaito, A., Tajiri, A., Hatano, M., Ogura, F., Nakagawa, M.: J. Amer. Chem. Soc. *98*, 7932 (1976)
32. Ito, O., Hatano, M.: J. Amer. Chem. Soc. *96*, 4375 (1974)
33. Kauzman, W. J., Walter, J. E., Eyring, H.: Chem. Rev. *26*, 339 (1940)
34. Moscowitz, A.: Adv. Chem. Phys. *4*, 67 (1962)
35. Weigang, O. E., Jr.: J. Chem. Phys *43*, 3607 (1965)
36. Howell, J. M.: J. Chem. Phys. *53*, 4152 (1970)
37. Pao, Y. H., Santry, D. P.: J. Amer. Chem. Soc. *88*, 4157 (1966)
38. Imamura, A., Hirano, T., Nagata, C., Tsuruta, T.: Bull. Chem. Jpn. *45*, 386 (1972)
39. Bouman, T. D., Voigt, B., Hansen, A. E.: J. Amer. Chem. Soc. *101*, 550 (1979)
40. Moffitt, W., Woodward, R. B., Moscowitz, A., Klyne, W., Djerassi, C.: J. Amer. Chem. Soc. *83*, 4013 (1961)
41. Höhn, E. G., Weigang, J., O. E.: J. Chem. Phys. *48*, 1127 (1968)
42. Klyne, W., Kirk, D. N.: "Fundamental Aspects and Recent Developments in Optical Rotatory Dispersion and Circular Dichroism", ed. by F. Ciardelli and P. Salvadori, Heyden & Son Ltd. New York (1973), p. 89
43. Snatzke, G., Snatzke-Zamonjska, F.: in Ref. 42, p. 109
44. Snatzke, G.: Tetrahedron Lett. 4275 (1972)
45. Kagan, H. B.: "Stereochemistry; Fundamentals and Methods", Vol. 2, "Determination of Configurations by Dipole Moments, CD or ORD", Georg Thieme Publishers, Stuttgart (1977)
46. Tinoco, Jr., I.: Adv. Chem. Phys. *4*, 113 (1962)
47. Mason, S. F.: Chem. Phys. Lett. *32*, 201 (1975)
48. Schipper, P. E., Nordén, B.: Chem. Phys. Lett. *67*, 99 (1979)
49. Craig, D. P., Power, E. A., Thirunamachandran, T.: Chem. Phys. Lett. *27*, 149 (1974)
50. Tajiri, A., Hirayama, H., Hatano, M.: Chem. Phys. Lett. *70*, 22 (1980)
51. Murakami, T., Hatano, M.: Inorg. Chem. *14*, 999 (1975)
52. Schipper, P. E.: J. Amer. Chem. Soc. *98*, 7938 (1976)
53. a) Andrews, L. J.: Chem. Rev. *54*, 713 (1954)
    b) Briegleb, G.: "Elektronen-Donator-Acceptor-Komplexe", Springer-Verlag, Berlin (1961)
    c) Mulliken, R. S., Person, W. B.: Ann. Rev. Phys. Chem. *13*, 107 (1962)
    d) Andrews, L. J., Keefer, R. M.: "Molecular Complexes in Organic Chemistry", Holden Day Inc., San Francisco (1964)
    e) Foster, R.: "Organic Charge-transfer Complexes", Academic Press, London (1969)
    f) Mulliken, R. S., Person, W. B.: "Molecular Complexes", Wiley-Interscience, New York (1969)
54. Briegleb, G., Kuball, H. G.: Angew. Chem. *76*, 228 (1964); Briegleb, G., Kuball, H. G., Henschel, K.: Z. Phys. Chem. *46*, 229 (1965)
55. Kuball, H. G., Henschel, K.: Z. Phys. Chem. *50*, 60 (1966); Briegleb, G., Kuball, H. G., Henschel, K., Euing, W.: Ber. Bunsenges. Phys. Chem. *76*, 101 (1972)
56. Tajiri, A., Gonohe, N., Hatano, M.: Ber. Bunsenges. Phys. Chem. *87*, 684 (1983)
57. Wellman, K. M., Laur, P. H. A., Briggs, W. S., Moscowitz, A., Djerassi, C.: J. Amer. Chem. Soc. *87*, 66 (1965)
58. Orgel, L. E., Mulliken, R. S.: J. Amer. Chem. Soc. *79*, 4839 (1957)
59. Saeva, F. D., Wysocki, J. J.: J. Amer. Chem. Soc. *93*, 5928 (1971)
60. Saeva, F. D.: J. Amer. Chem. Soc. *94*, 5135 (1972)
61. Saeva, F. D., Olin, G. R.: J. Amer. Chem. Soc. *95*, 7882 (1973)
62. Saeva, F. D., Sharpe, P. E., Olin, G. R.: J. Amer. Chem. Soc. *95*, 7656 (1973)
63. Sackmann, E., Voss, J.: Chem. Phys. Lett. *14*, 528 (1972)

64. Sackmann, E., Möhwald, H.: J. Chem. Phys. *58*, 5407 (1973)
65. de Vries, H.: Acta Crystallogr. *4*, 214 (1951)
66. Mason, S. F., Peacock, R. D.: J. Chem. Soc. Chem. Commun. 712 (1973)
67. Sato, y., Hatano, M.: Makromol. Chem. *183*, 971 (1982)
68. Sato, Y., Tajiri, A., Hatano, M.: Makromol. Chem. *183*, 989 (1982)
69. Sato, Y., Hatano, M.: Makromol. Chem. *183*, 997 (1982)
70. Zimmerman, S., Pheiffer, B.: J. Mol. Biol. *142*, 315 (1980)
71. Chen, C. Y., Pheiffer, B., Zimmerman, S., Hanlon, S.: Biochemistry *22*, 4746 (1983)
72. Hanlon, S., Brudno, S., Wu, T. T., Wolf, B.: Biochemistry *14*, 1648 (1975)
73. Behe, M., Felsenfeld, G.: Proc. Natl. Acad. Sci., U.S.A. *78*, 1619 (1981)
74. Patel, D. J., Cannel, L. L., Pohl, F. M.: Proc. Natl. Acad. Sci., U.S.A. *76*, 2508 (1979)
75. Wang, A. H.-J., Quigley, G. J., Kolpak, F. J., Crawford, J. L., van Boom, J. H., van der Marel, G., Rich, A.: Nature (London) *282*, 680 (1979)
76. Pohl, F. M., Jovin, T. M.: J. Mol. Biol. *67*, 375 (1972)
77. McGhee, J. D., Ginder, D. D.: Nature (London), *280*, 419 (1979)
78. Chen, C.-W., Cohen, J. C., Behe, M.: Biochemistry *22*, 2136 (1983)
79. Neville, D. M., Bradeley, D. F.: Biochim. Biophys. Acta *50*, 397 (1961)
80. Blake, A., Peacock, A. R.: Biopolymers *4*, 1091 (1966); Dalgleich, D. G., Peacock, A. R., Fey, G., Harvey, M.: Biopolymers *10*, 1853 (1971)
81. Schipper, E., Nordén, B., Tjerneld, F.: Chem. Phys. Lett. *70*, 17 (1980)
82. Sobell, H., Jain, S. C.: J. Mol. Biol. *114*, 301, 317, 333 (1977); ibid. *68*, 21 (1972)
83. Gangain, B., Markovits, J., Le Peeq, J.-B., Roques, B.: Biochemistry *20*, 3035 (1981)
84. Chen, C., Cohen, J. S., Zador, A.: J. Biochem. Biophys. Methods *5*, 293 (1981)
85. Chen, C., Knop, R. H., Cohen, J. S.: Biochemistry *22*, 5468 (1983)
86. Wells, B. D., Yang, J. T.: Biochemistry *13*, 1317 (1974)
87. McIntosh, L. P., Grieger, I., Eckstein, F., Zarling, D. A., van de Sande, J. H., Jovin, T. M.: Nature (London) *304*, 83 (1983)
88. Wang, A., Quigley, G., Kolpak, F., van der Marel, G., van Boom, J., Rich, A.: Science (Washington, D.C.) *211*, 171 (1981)
89. Hingery, B., Broyde, S.: Biochemistry *21*, 3243 (1982)
90. Holler, E., Holmquist, B., Valee, B. L., Taneja, K., Zamecnik, P.: Biochemistry *22*, 4924 (1983)
90a Greenfield, N., Fasman, G. D.: Biochemistry *8*, 4108 (1969)
91. Sexena, V. P., Wetlaufer, D. B.: Proc. Natl. Acad. Sci., U.S.A. *68*, 969 (1971)
92. Bannister, W. H., Bannister, J. V.: Int. J. Biochem. *5*, 673 (1974)
93. Chang, C. T., Wu, C-S. C., Yang, J. T.: Analyt. Biochem. *91*, 13 (1978)
94. Brahms, S., Brahms, J.: J. Mol. Biol. *138*, 149 (1980)
95. Hennessey, Jr., J. P., Johnson, Jr., W. C.: Biochemistry *20*, 1085 (1981)
96. Manavalan, P., Johnson, Jr., W. C.: Nature (London) *305*, 831 (1983)
97. Schechter, E., Blout, R. R.: Proc. Natl. Acad. Sci. U.S.A. *51*, 695, 794 (1963)
98. Shiraki, M., Imahori, K.: Sci. Pap. College Gen. Educ., Univ. Tokyo *19*, 151 (1969)
99. Woody, R. W.: J. Chem. Phys. *49*, 4797 (1968); Tetrahedron *29*, 1273 (1973)
100. Goux, W. J., Hooker, Jr., T. M.: J. Amer. Chem. Soc. *102*, 7080 (1980)
101. Konishi, Y., van Nispen, J. W., Davenport, G., Scheraga, H. A.: Macromolecules *10*, 1264 (1977)
102. Hatano, M., Yoneyama, M.: J. Amer. Chem. Soc. *92*, 1392 (1970)
103. Chen, Y-H., Yang, J. T., Chan, K. H.: Biochemistry *13*, 3350 (1974)
104. Pocker, Y., Biswas, S. B.: Biochemistry *19*, 5043 (1980)
105. Chou, P. Y., Fasman, G. D.: Biochemistry *13*, 222 (1974)
106. Chou, P. Y., Fasman, G. D.: Annu. Rev. Biochem. *47*, 251 (1978)
107. Shikama, K., Suzuki, T., Sugawara, Y., Katagiri, T., Takagi, T., Hatano, M.: Biochim. Biophys. Acta *701*, 138 (1982)
108. Hennessey, J. P., Johnson, Jr., W. C.: Biochemistry *20*, 1085 (1981)
109. Chou, P. Y., Fasman, G. D.: J. Mol. Biol. *115*, 135 (1977)
110. Provencher, S. W., Glöckner, J.: Biochemistry *20*, 33 (1981)
111. Tanford, C.: Advances in Protein Chemistry *23*, 121 (1968); ibid. *24*, 1 (1970)
112. Kuwajima, K., Sugai, S.: Biophys. Chem. *8*, 247 (1978)
113. Nitta, K., Kita, N., Kuwajima, K., Sugai, S.: Biochim. Biophys. acta *490*, 200 (1977)

114. Nozaka, M., Kuwajima, K., Nitta, K., Sugai, S.: Biochemistry *17*, 3753 (1978); ibid. *24*, 874 (1985)
115. Hauschka, P. V., Carr, S. A.: Biochemistry 21, 2538 (1982)
116. Kakiuchi, S., Hidaka, S., Meansm, A. R.: "Calmodulin and Intracellular Ca^{2+} Receptors", Plenum Press, New York and London (1982)
117. Crouch, T. H., Klee, C. B.: Biochemistry *19*, 3692 (1980)
118. Shimizu, T., Hatano, M.: FEBS Lett. *160*, 182 (1983); Shimizu, T., Hatano, M.: Biochemistry *23*, 6403 (1984)
119. Shimizu, T., Hatano, M., Muto, Y., Nozawa, Y.: FEBS Lett. *166*, 373 (1984)
120. Shimizu, T., Hatano, M., Nagao, S., Nozawa, Y.: Biochem. Biophys. Res. Commun. *106*, 1112 (1982); Shimizu, T., Hatano, M.: Inorg. Chem. *24*, 2003 (1985)
121. Takagi, S., Nomori, H., Hatano, M.: Chem. Lett. 611 (1974)
122. Nomori, H., Tsuchihashi, N., Takagi, S., Hatano, M.: Bull. Chem. Soc. Jpn. *48*, 2522 (1975)
123. Robinson, C.: Mol. Crystr. *1*, 467 (1966)
124. Hatano, M., Enomoto, T., Ito, I., Yoneyama, M.: Bull. Chem. Soc. Jpn. *46*, 3698 (1973)
125. Ueno, A., Toda, F., Iwakura, Y.: Biopolymers *13*, 1213 (1974)
126. Enomoto, T., Nomori, H., Hatano, M.: Chem. Lett. 1289 (1974)
127. Konishi, Y., Hatano, M.: J. Polymer Sci., Polymer Lett. Ed. *14*, 351 (1976)
128. Yoshikawa, M., Nomori, H., Hatano, M.: Makromol. Chem., *179*, 2397 (1978)
129. Ueno, A., Ishigura, T., Toda, F., Uno, K., Iwakura, Y.: Biopolymers *14*, 353 (1975)
130. Ueno, A., Nohara, M., Toda, F., Uno, K., Iwakura, Y.: J. Polymer Sci., Polymer Lett. Ed. *13*, 2751 (1975)
131. Ueno, A., Osa, T., Toda, F.: J. Polymer Sci., Polymer Lett. Ed. *16*, 539 (1978)
132. Tsuchihashi, N., Enomoto, T., Tanikawa, A., Tajiri, A., Hatano, M.: Makromol. Chem. *176*, 2833 (1975); Chapoy, L. L., Biddle, D., Halstrom, J., Kovacs, K., Brunfeldt, K., Qasim, M. A., Christensen, T.: Macromolecules *16*, 181 (1983)
133. Tajiri, A., Hatano, M.: to be published elsewhere.
134. Sage, H. J., Fasman, G. D.: Biochemistry 5, 286 (1966)
135. Auer, H. E., Doty, P.: Biochemistry 5, 1708 (1966)
136. Urry, D. W.: Annun. Rev. Phys. Chem. *19*, 477 (1968)
137. Shiraki, M., Imahori, K.: Sci. Pap. College Gen. Educ., Univ. Tokyo *16*, 215 (1966)
138. Damle, V. N.: Biopolymers *9*, 937 (1970)
139. Quadrifoglio, F., Ius, A., Crescenzi, V.: Makromol. Chem. *136*, 241 (1970)
140. Cosani, A., Peggion, E., Verdini, A. S., Terbojevich, M.: Biopolymers *6*, 963 (1968)
141. Peggion, E., Cosani, A., Verdini, A., Del Pra, A., Mammi, M.: Biopolymers *6*, 1477 (1968)
142. Goodman, M., Kossoy, A.: J. Amer. Chem. Soc. *88*, 5010 (1966)
143. Goodman, M., Peggion, E.: Biochemistry *6*, 1533 (1967)
144. Peggion, E., Cosani, E., Palumbo, M., Terbofevich, M., Goodman, M.: Biopolymers *15*, 2227 (1976)
145. Goodman, M., Tiniolo, C., Peggion, E.: Biopolymers *6*, 1691 (1968)
146. Sisido, M., Egusa, S., Imanishi, Y.: J. Amer. Chem. Soc. *105*, 1041, 4077 (1983)
147. Tajiri, A., Hatano, M.: Chem. Phys. Lett. *34*, 29 (1975)
148. Blout, E. R., Stryer, L.: Proc. Natl. Acad. Sci., U.S.A. *45*, 1591 (1959); Idem.: J. Amer. Chem. Soc. *83*, 1411 (1961)
149. Hatano, M., Yoneyama, M., Sato, Y.: Biopolymers *12*, 895 (1973)
150. Yamamoto, S., Nozawa, T., Hatano, M.: Polymer (London) *15*, 330 (1974)
151. Hatano, M., Yoneyama, M., Sato, Y., Kawamura, Y.: Biopolymers *12*, 2423 (1973)
152. Sato, Y., Hatano, M., Yoneyama, M.: Bull. Chem. Soc. Jpn. *46*, 1980 (1973)
153. Sato, Y., Hatano, M.: Bull. Chem. Soc. Jpn. *46*, 3339 (1973)
154. Ikeda, S., Imae, T.: Biopolymers *10*, 1743 (1971)
155. Yamamoto, H., Nakazawa, A.: Chem. Lett. 47 (1983)
156. Yamamoto, H., Nakazawa, A., Hayakawa, T.: Polymer Sci., Polymer Lett. Ed. *21*, 131 (1983)
157. Yamamoto, H., Nakazawa, A.: Bull. Chem. Soc. Jpn. *56*, 2535 (1983)
158. Shimizu, T., Nozawa, T., Hatano, M., Satake, H., Imai, Y., Hashimoto, C., Sato, R.: Biochim. Biophys. Acta *579*, 122 (1979)
159. Imai, Y.: J. Biochem. *92*, 57 (1982)
160. Levin, R. M., Weiss, B.: Mol. Pharmacol. *13*, 690 (1977)

161. Bromstrom, C. O., Bromstrom, M. A., Wolff, D. J.: in "Calcium Regulation by Calcium Antagonists", Rahwan, R. G., Witiak, D. T. eds., p. 89–103, American Chemical Society, Washington, D.C. (1982)
162. Shimizu, H., Kaito, A., Hatano, M.: J. Amer. Chem. Soc. *104*, 7059 (1982); Idem., Bull. Chem. Soc. Jpn. *52*, 2678 (1979); ibid. *54*, 513 (1981)
163. Collet, A., Gottarelli, G.: J. Amer. Chem. Soc. *104*, 7383 (1982)
164. Buss, V., Kolster, K., Winger, U., Simon, L.: J. Amer. Chem. Soc. *106*, 4621 (1984)
165. Kaito, A., Hatano, M.: J. Amer. Chem. Soc. *100*, 4037 (1978)
166. Thulstrup, E. W.: "Linear and Magnetic Circular Dichroism of Planar Organic Molecules", Springer-Verlag, Heidelberg (1980)
167. Schipper, P. E., Rodger, A.: J. Amer. Chem. Soc. *105*, 4541 (1983)
168. Taillander, E., Taboury, J. A., Adam, S., Liquier, J.: Biochemistry *23*, 5703 (1984)
169. Russell, W. C., Precious, B., Martin, S. R., Bayley, P. M.: EMBO J. *2*, 1647 (1983)
170. Miller, F. D., Dixon, G. H., Rattner, J. B., van de Sande, J. H.: Biochemistry *24*, 102 (1985)
171. Limn, W., Vesugi, S., Ikehara, M., Miles, H. T.: Biochemistry *22*, 4217 (1983)
172. Cleare, M. J., Hydes, P. C.: Metal Ions in Biological Systems, Vol. 11, 1 (1980)
173. Chottard, J. C., Girault, J. P., Chottard, G., Lallemand, J. Y., Mansuy, D. J.: J. Amer. Soc. *102*, 5565 (1980); Gitraut, J. P., Chottard, G., Lallemand, J. Y., Chlttard, J. C.: Biochemistry *21*, 1352 (1982)
174. Tullius, T. D., Lippard, S. J.: Proc. Natl. Acad. Sci., U.S.A. *79*, 3489 (1982)
175. Caradonna, J. P., Lippard, S. J., Gait, M. J., Singh, M.: J. Amer. Chem. Soc. *104*, 5793 (1982)
176. Inagaki, K., Kasuya, K., Kidani, Y.: Chem. Lett. *1984*, 171
177. Barton, J. K., Danishefsky, A. T., Goldberg, J. M.: J. Amer. Chem. Soc. *106*, 2172 (1984)
178. Barton, J. K., Lolis, E.: J. Amer. Chem. Soc. *107*, 708 (1985)
179. Barton, J. K., Danishefsky, A. T., Goldberg, J. M.: J. Amer. Chem. Soc. *106*, 2172 (1984)
180. Tsuruo, T., Iida, H., Tsukagoshi, S., Sakurai, Y.: Jpn. J. Cancer Res. *7*, 151 (1980)
181. Goodwin, D. C., Brahms, J.: Nucleic Acids Res. *5*, 835 (1978)
182. Bram, S.: J. Mol. Biol. *58*, 277 (1971)
183. Cowman, M. K., Fasman, G. D.: Biochemistry *19*, 532 (1980)
184. Jordrano, J., Montero, F., Palacián, E.: Biochemistry *23*, 4285 (1984)
185. Wu, H.-Y., Behe, M. J.: Proc. Natl. Acad. Sci., U.S.A. *81*, 7284 (1984)
186. Kyte, J., Doolittle, R. F.: J. Mol. Biol. *157*, 105 (1982)
187. Tarr, G. E., Black, S. D., Fujita, V. S., Coon, M. J.: Proc. Natl. Acad. Sci., U.S.A. *80*, 6552 (1983)
188. Widger, W. R., Cramer, W. A., Herrmann, R. G., Trebst, A.: Proc. Natl. Acad. Sci., U.S.A. *81*, 674 (1984)
189. Williams, J. C., Steiner, L. A., Feher, G., Simon, M. I.: Proc. Natl. Acad. Sci., U.S.A. *81*, 7303 (1984)
190. Youvan, D. C., Ismail, S.: Proc. Natl. Acad. Sci., U.S.A. *82*, 58 (1985)
191. Nozawa, T., Ohta, M., Hatano, M., Hayashi, H., Shimada, K.: Chem. Lett. *1985*, 343
192. Nozawa, T., Hatano, M.: to be submitted
193. Hsu, W., Woody, R. W.: J. Amer. Chem. Soc. *93*, 3515 (1971)
194. Tinoco, Jr., I., Advances in Chem. Phys. *4*, 113 (1962); Woody, R. W., Tinoco, Jr., I.: Chem. Phys. *46*, 4527 (1967)
195. Perutz, M. F., Muirhead, H., Cox, J. M., Goaman, L. C. G.: Nature (London) *219*, 131 (1968)
196. Urry, D. W., Pellegrew, J. W.: J. Amer. Chem. Soc. *89*, 5257 (1967)
197. Urry, D. W.: J. Amer. Chem. Soc. *89*, 4190 (1967)
198. Pande, A. J., MacDonald, L. H., Myer, Y. P.: Biophys. J. *15*, 286a (1975)
199. Okuyama, K., Murakami, T., Nozawa, T., Hatano, M.: Chem. Lett. *1982*, 111
200. Myer, Y. P., Pande, A. J.: in "The Porphyrins", Dolphin, D., ed., Vol. 3; Physical Chemistry, Part A, p. 271, Academic Press, New York (1978)
201. Wright, K. A., Boxer, S. G.: Biochemistry *20*, 7546 (1981)
202. Kuki, A., Boxer, S. G.: Biochemistry *22*, 2923 (1983)
203. Sauer, K.: Acc. Chem. Res. *11*, 257 (1978)
204. Sauer, K., Austin, L. A.: Biochemistry *17*, 2011 (1978)
205. Miyazaki, T., Morita, S., Hatano, M., Nozawa, T.: J. Biochem. *86*, 1411 (1979)
206. Bolt, J. D., Sauer, K., Shiozawa, J. A., Drews, G.: Biochim. Biophys. Acta *635*, 535 (1981)

207. Hayashi, H., Nozawa, T., Hatano, M., Morita, S.: J. Biochem. *89*, 1853 (1981)
208. Hayashi, H., Nozawa, T., Hatano, M., Morita, S.: J. Biochem. *91*, 1029 (1982)
209. Cogdell, R. J., Thornber, J. P.: "Chlorophyll Organization and Energy Transfer is Photosynthesis", Ciba Foundation Symposium No. 61, Exerpta Media, Amsterdam, Oxford & New York, 1979, P. 61.
210. Tadros, M. T., Suter, F., Seydewitz, H. H., Witt, I., Zuber, H., Drews, G.: Eur. J. Biochem. *138*, 209 (1984)
211. Bruinisholz, R. A., Guendet, P. A., Theiler, R., Zuber, H.: FEBS Lett. *129*, 150 (1981)
212. Gogel, G. E., Parkes, P. S., Loach, P. A., Brunisholz, R. A., Zuber, H.; Biochem. Biophys. Acta *746*, 32 (1983)
213. Youvan, D. C., Alberti, M., Begusch, H., Bylina, E. J., Hearst, J. E.: Proc. Natl. Acad. Sci., U.S.A. *81*, 189 (1984)
214. Bucks, R. R., Boxer, S. G.: J. Amer. Chem. Soc. *104*, 340 (1982)
215. Hayashi, H., Hamaguchi, H., Tasumi, M.: Chem. Lett. *1983*, 1857
216. Nozawa, T., Nishimura, M., Hatano, M., Hayashi, H., Shimada, K.: Biochemistry. *24*, 1890 (1985)
217. Sigel, H., ed.: "Metal Ions in Biological Systems", Vol. 1–Vol. 16, Marcel Dekker, Inc., New York and Basel (1973–1983); Hill, H. A. O., ed.: "Inorganic Biochemistry", Vol. 1–Vol. 3, The Royal Society of Chemistry, London (1979–1982); Theil, E., Eichhorn, G., Marzilli, L., eds.: "Advances in Inorganic Biochemistry", Vol. 1–Vol. 5, Elsevier Science Publ. Co., Inc., New York and Amsterdam (1979–1983)
218. Verheij, H. M., Volwerk, J. J., Jansen, E. H. J. M., Puyk, W. C., Dijkstra, B. W., Drenth, J., de Haas, G. H.: Biochemistry *19*, 743 (1980)
219. Tucker, P. W., Hazen, E. E., Cotton, F. A.: Mol. Cell Biochem. *23*, 67 (1979)
220. Matthews, B. M., Weaver, L. H.: Biochemistry *13*, 1719 (1974)
221. Roche, R. S., Voordouw, G.: in "Calcium Binding Proteins and Calcium Function", Wasserman, R. H., ed., p. 38, Elsevier Science Publ. Co., Inc., New York and Amsterdam (1977)
222. Klee, C. B.: Biochemistry *16*, 1017 (1977)
223. Kretsinger, R. H.: in Ref. 221, p. 63
224. Morpeth, F. F., Massey, V.: Biochemistry *21*, 1318 (1982)
225. Falk, M. C., Bethune, J. L., Valee, B. L.: Biochemistry *21*, 1471 (1982)
226. Forsén, S., Lindeman, B.: Ann. Rep. NMR. Spectr. *11 A*, 183 (1981)
227. Dunbar, J. C., Holmquist, B.: Biochemistry *23*, 4330 (1984)
228. Nitta, K., Segawa, T., Kuwajima, K., Sugai, S.: Biopolymers *16*, 703 (1977)
229. Bayley, P. M.: Progr. Biophys. Methods Biolog. *37*, 149 (1979)
230. Segawa, T., Kuwajima, K., Sugai, S.: Biochim. Biophys. Acta *668*, 89 (1981)
231. Hasumi, H.: Biochim. Biophys. Acta *626*, 265 (1980)
232. Anson, M., Martin, S. R., Bayley, P.: Rev. Sci. Instrum. *48*, 953 (1977)
233. Hatano, M., Nozawa, T., Murakami, T., Yamamoto, T., Shigehisa, M., Kimura, M., Takakuwa, N., Sakayanagi, N., Yano, T., Watanabe, A.: Rev. Sci. Instrum. *52*, 1311 (1981)
234. Johnson, Jr., W. C.: Rev. Sci. Instrum. *42*, 1283 (1971); Johnson, Jr., W. C.: Ann. Rev. Phys. Chem. *29*, 93 (1978)
235. Nelson, R. G., Johnson, Jr., W. C.: J. Amer. Chem. Soc. *94*, 3343 (1972); ibid. *98*, 4296 (1976)
236. Stipanovic, A. J., Stevens, E. S.: in "Solution Properties of Polysaccharides", Brant, D. A., ed., American Chemical Society Symposium Series, No. 150, American Chemical Society, Washington (1981), p. 303
237. Nishida, K., Iwasaki, A.: Kolloid Z. u. Z. Polymere *251*, 136 (1973)
238. Chakrabarti, B., Balasz, E. E.: Biochem. Biophys. Res. Commun. *52*, 1170 (1973)
239. Ogawa, K., Hatano, M.: Carbohydrate Res. *67*, 527 (1978)
240. Buffington, L. A., Pysh, E. S., Chakrabarti, B., Balazs, E. E.: J. Amer. Chem. Soc. *99*, 1730 (1977)
241. Park, J. W., Chakrabarti, B.: Biopolymers *17*, 1323 (1978)
242. Park, J. W., Chakrabarti, B.: Biochim. Biophys. Acta *541*, 263 (1978)
243. Buffington, L. A., Stevens, E. S.: J. Amer. Chem. Soc. *101*, 5159 (1979)
244. Allerhand, A., Berman, E.: J. Amer. Chem. Soc. *106*, 2400 (1984); ibid. *106*, 2412 (1984)
245. Berman, E., Allerhand, A., Devries, A. L.: J. Biol. Chem. *255*, 4407 (1980)
246. Liu, H.-W., Nakanishi, K.: J. Amer. Chem. Soc. *104*, 1178 (1982)

247. Snyder, P. A., Rowe, E. M.: Nuclear Instruments and Methods *172*, 345 (1980)
248. Snyder, P. A., Lund, P. A., Schatz, P. N., Rowe, E. M.: Chem. Phys. Lett. *82*, 546 (1981)
249. Sackmann, E., Voss, J.: Chem. Phys. Lett. *14*, 528 (1972)
250. Shindo, Y., Ohmi, Y.: J. Amer. Chem. Soc. *107*, 91 (1985)
251. Tachibana, T., Mori, T., Hori, K.: Nature (London) *278*, 578 (1979)
252. Sakamoto, K., Yoshida, R., Hatano, M., Tachibana, T.: J. Amer. Chem. Soc. *100*, 6898 (1978)
253. Christiansen, C.: Justus Liebigs Ann. Chem. *23*, 289 (1884)
254. Sato, K.: Bull. Chem. Soc. Jpn. *17*, 31 (1942)
255. Holmes, H. N., Cameron, D. H.: J. Amer. Chem. Soc. *44*, 71 (1922)
256. Kano, M., Tajiri, A., Hatano, M.: in preparation
257. Charvolin, J., Deloche, B., in "The Molecular Physics of Liquid Crystals", Luckhurst, G. R., Gray, G. W., eds., Academic Press, New York, 1979, p. 343
258. Okahata, Y., Kunitake, T.: J. Amer. Chem. Soc. *101*, 5231 (1979)
259. Kunitake, T., Okahata, Y.: J. Amer. Chem. Soc. *102*, 549 (1980)
260. Okahata, Y., Kunitake, T.: Ber. Bunsenges. Phys. Chem. *84*, 550 (1980)
261. Kunitake, T., Nakashima, N., Simomura, M., Okahata, Y., Kano, K., Ogawa, T.: J. Amer. Chem. Soc. *102*, 6642 (1980)
262. Jonansson, L. B.-Å., Davidson, Å, Lindblom, G., Nordén, B.: J. Phys. Chem. *82*, 2604 (1978)
263. Nakashima, N., Fukushima, M., Kunitake, T.: Chem. Lett. *1981*, 1207
264. Kunitake, T., Nakashima, N., Morimitsu, K.: Chem. Lett. *1980*, 1347
265. Nakashima, N., Morimitsu, K., Kunitake, T.: Bull. Chem. Soc. Jpn. *57*, 3253 (1984)
266. Nakashima, N., Asakuma, S., Kunitake, T.: J. Amer. Chem. Soc. *107*, 509 (1985); Nakashima, N., Asakuma, S., Kim, J.-M., Kunitake, T.: Chem. Lett. *1984*, 1709
267. Lin, K.-C., Weis, R. M., McConnell, H. M.: Nature (London) *296*, 164 (1982)
268. Tachibana, T., Mori, K.: J. Colloid Interface Sci. *61*, 398 (1977)
269. Tachibana, T., Kayama, K., Takeno, H.: Bull. Chem. Soc. Jpn. *42*, 3422 (1969)
270. Uzu, K., Sugiura, T.: J. Colloid Interface Sci. *51*, 346 (1975)
271. Tachibana, T., Yoshizumi, T., Hori, K.: Bull. Chem. Soc. Jpn. *52*, 34 (1979)
272. Stewart, M. V., Arnett, E. M.: Topics Spectrochem. *13*, 195 (1982)
273. Kuball, H.-G., Karstens, T.: Chem. Phys. *12*, 1 (1976)
274. Kuball, H.-G., Altschuh, J., Kulbach, R.: Helv. Chim. Acta *61*, 571 (1978)
275. Kuball, H.-G., Acimis, M., Altschuh, J.: J. Amer. Chem. Soc. *101*, 20 (1979)
276. Kuball, H.-G., Altschuh, J.: Chem. Phys. Lett. *87*, 599 (1982)
277. Bustamante, C., Tinoco, Jr., I., Maestre, M. F.: Proc. Natl. Acad. Sci., U.S.A. *80*, 3568 (1983)
278. Mao, D., Wallace, B. A.: Biochemistry *23*, 2667 (1984)
279. Reich, C., Maestre, M. F., Edmondson, S., Gray, D. M.: Biochemistry *19*, 5208 (1980)
280. Tsujimura, K., Konno, T., Meguro, H., Hatano, M., Murakami, T., Kashiwabara, K., Saito, K., Kondo, Y., Suzuki, T. M.: Analyt. Biochem. *81*, 167 (1977)
281. Konno, T., Meguro, H., Murakami, T., Hatano, M.: Chem. Lett. *1981*, 953
282. Meyer, B.: "Low Temperature Spectroscopy", American Elsevier Publishing Co., Inc., New York, 1971, p. 204
283. Nozawa, T., Shimizu, T., Hatano, M., Shimada, H., Iizuka, T., Ishimura, Y.: Biochim. Biophys. Acta *534*, 285 (1978)
284. Diem, M., Gotkin, P. J., Kupfer, J. M., Nafie, L. A.: J. Amer. Chem. Soc. *100*, 5644 (1978)
285. Nafie, L. A., Diem, M., Vidrine, D. W.: J. Amer. Chem. Soc. *101*, 496 (1979)
286. Bilardon, M., Badoz, J.: Compt. Rend. Acad. Sci. *B262*, 1672 (1966); ibid. *B263*, 26 (1966)
287. unpublished data
288. Su, C. N., Heinz, V. J., Keiderling, T. A.: Chem. Phys. Lett. *73*, 157 (1980)
289. Saito, Y., Wada, A.: Biopolymers *22*, 2123 (1983)
290. Buckingham, A. D., Stephens, P. J.: Annu. Rev. Phys. Chem. *17*, 399 (1966)
291. Schatz, P. N., McCaffery, A. J.: Quart. Rev. Chem. Soc. *23*, 552 (1969)
292. Caldwell, D., Thorne, J. M., Eyring, H.: Annu. Rev. Phys. Chem. *22*, 259 (1971)
293. Stephens, P. J.: Annu. Rev. Phys. Chem. *25*, 201 (1974)
294. Stephens, P. J., Mowery, R. L., Schatz, P. N.: J. Chem. Phys. *55*, 224 (1971)
295. Stephens, P. J., Schatz, P. N., Ritchie, A. B., McCaffery, A. J.: J. Chem. Phys. *48*, 132 (1968)
296. Kaito, A., Tajiri, A., Hatano, M.: Chem. Phys. Lett. *28*, 197 (1974)
296. Kaito, A., Tajiri, A., Hatano, M.: J. Amer. Chem. Soc. *97*, 5059 (1975)

297. Kaito, A., Tajiri, A., Hatano, M.: J. Amer. Chem. Soc. *98*, 384 (1976)
298. Kaito, A., Hatano, M.: J. Amer. Chem. Soc. *100*, 2034 (1978)
299. Kaito, A., Tajiri, A., Hatano, M., Ogura, F., Nakagawa, M.: J. Amer. Chem. Soc. *98*, 7932 (1976)
300. Kaito, A., Hatano, M., Tajiri, A.: J. Amer. Chem. Soc. *99*, 5241 (1977)
301. Kaito, A., Tajiri, A., Hatano, M.: Bull. Chem. Soc. Jpn. *49*, 2207 (1976)
302. Kaito, A., Hatano, M., Ueda, T., Shibuya, S.: Bull. Chem. Soc. Jpn. *53*, 3073 (1980)
303. Kaito, A., Hatano, M.: Bull. Chem. Soc. Jpn. *53*, 3064 (1980)
304. Kaito, A., Hatano, M.: Bull. Chem. Soc. Jpn. *53*, 3069 (1980)
305. Tajiri, A., Hatano, M., Nakazawa, T., Murata, I.: Chem. Phys. Lett. *76*, 490 (1980)
306. Igarashi, N., Tajiri, A., Hatano, M.: Bull. Chem. Soc. Jpn. *54*, 1511 (1981)
307. Tajiri, A., Hatano, M., Oda, M.: Chem. Phys. Lett. *78*, 112 (1981)
308. Tajiri, A., Hatano, M., Morita, T., Saito, M., Takase, K.: Chem. Phys. Lett. *83*, 101 (1981)
309. Tajiri, A., Hatano, M., Toda, T., Shimazaki, N., Mukai, T.: Chem. Phys. Lett. *81*, 251 (1981)
310. Uchimura, H., Tajiri, A., Hafano, M.: Bull. Chem. Soc. Jpn. *54*, 3279 (1981)
311. Tajiri, A., Hatano, M., Murata, I.: Ber. Bunsenges. Phys. Chem. *86*, 228 (1982)
312. Fukuda, M., Tajiri, A., Oda, M., Hatano, M.: Bull. Chem. Soc. Jpn. *56*, 592 (1983)
313. Tajiri, A., Hatano, M., Yamamoto, K., Murata, I.: Chem. Phys. Lett. *91*, 433 (1982)
314. Tajiri, A., Hatano, M., Takahashi, K.: Chem. Phys. Lett., Chem. Phys. Lett. *98*, 290 (1983)
315. Tajiri, A., Fukuda, M., Hatano, M., Morita, T., Takase, K.: Angew. Chem. *95*, 911 (1983)
316. Tajiri, A., Yamamoto-Igarashi, N., Hatano, M., Ueda, T.: Heterocycles *22*, 2053 (1984)
317. Tajiri, A., Fukuda, M., Hatano, M., Yasunami, M., Takagi, A., Takase, K.: Chem. Phys. Lett. *108*, 378 (1984)
318. Tajiri, A., Hatano, M., Asao, T., Morita, N.: Angew. Chem., in press (1985)

Author Index Volumes 1–76

Allegra, G. and *Bassi*, I. W.: Isomorphism in Synthetic Macromolecular Systems. Vol. 6, pp. 549–574.
Andrews, E. H.: Molecular Fracture in Polymers. Vol. 27, pp. 1–66.
Anufrieva, E. V. and *Gotlib*, Yu. Ya.: Investigation of Polymers in Solution by Polarized Luminescence. Vol. 40, pp. 1–68.
Apicella, A. and *Nicolais*, L.: Effect of Water on the Properties of Epoxy Matrix and Composite. Vol. 72, pp. 69–78.
Apicella, A., *Nicolais*, L. and *de Cataldis*, C.: Characterization of the Morphological Fine Structure of Commercial Thermosetting Resins Through Hygrothermal Experiments. Vol. 66, pp. 189–208.
Argon, A. S., *Cohen*, R. E.. *Gebizlioglu*, O. S. and *Schwier*, C.: Crazing in Block Copolymers and Blends. Vol. 52/53, pp. 275–334
Arridge, R. C. and *Barham*, P. J.: Polymer Elasticity. Discrete and Continuum Models. Vol. 46, pp. 67–117.
Aseeva, R. M., *Zaikov*, G. E.: Flammability of Polymeric Materials. Vol. 70, pp. 171–230.
Ayrey, G.: The Use of Isotopes in Polymer Analysis. Vol. 6, pp. 128–148.

Bässler, H.: Photopolymerization of Diacetylenes. Vol. 63, pp. 1–48.
Baldwin, R. L.: Sedimentation of High Polymers. Vol. 1, pp. 451–511.
Balta-Calleja, F. J.: Microhardness Relating to Crystalline Polymers. Vol. 66, pp. 117–148.
Barton, J. M.: The Application of Differential Scanning Calorimetry (DSC) to the Study of Epoxy Resins Curing Reactions. Vol. 72, pp. 111–154.
Basedow, A. M. and *Ebert*, K.: Ultrasonic Degradation of Polymers in Solution. Vol. 22, pp. 83–148.
Batz, H.-G.: Polymeric Drugs. Vol. 23, pp. 25–53.
Bell, J. P. see *Schmidt*, R. G.: Vol. 75, pp. 33–72.
Bekturov, E. A. and *Bimendina*, L. A.: Interpolymer Complexes. Vol. 41, pp. 99–147.
Bergsma, F. and *Kruissink*, Ch. A.: Ion-Exchange Membranes. Vol. 2, pp. 307–362.
Berlin, Al. Al., *Volfson*, S. A., and *Enikolopian*, N. S.: Kinetics of Polymerization Processes. Vol. 38, pp. 89–140.
Berry, G. C. and *Fox*, T. G.: The Viscosity of Polymers and Their Concentrated Solutions. Vol. 5, pp. 261–357.
Bevington, J. C.: Isotopic Methods in Polymer Chemistry. Vol. 2, pp. 1–17.
Bhuiyan, A. L.: Some Problems Encountered with Degradation Mechanisms of Addition Polymers. Vol. 47, pp. 1–65.
Bird, R. B., *Warner*, Jr., H. R., and *Evans*, D. C.: Kinetik Theory and Rheology of Dumbbell Suspensions with Brownian Motion. Vol. 8, pp. 1–90.
Biswas, M. and *Maity*, C.: Molecular Sieves as Polymerization Catalysts. Vol. 31, pp. 47–88.
Biswas, M., *Packirisamy*, S.: Synthetic Ion-Exchange Resins. Vol. 70, pp. 71–118.
Block, H.: The Nature and Application of Electrical Phenomena in Polymers. Vol. 33, pp. 93–167.
Bodor, G.: X-ray Line Shape Analysis. A. Means for the Characterization of Crystalline Polymers. Vol. 67, pp. 165–194.
Böhm, L. L., *Chmeliř*, M., *Löhr*, G., *Schmitt*, B. J. and *Schulz*, G. V.: Zustände und Reaktionen des Carbanions bei der anionischen Polymerisation des Styrols. Vol. 9, pp. 1–45.

Bovey, F. A. and *Tiers, G. V. D.:* The High Resolution Nuclear Magnetic Resonance Spectroscopy of Polymers. Vol. 3, pp. 139–195.

Braun, J.-M. and *Guillet, J. E.:* Study of Polymers by Inverse Gas Chromatography. Vol. 21, pp. 107–145.

Breitenbach, J. W., Olaj, O. F. und *Sommer, F.:* Polymerisationsanregung durch Elektrolyse. Vol. 9, pp. 47–227.

Bresler, S. E. and *Kazbekov, E. N.:* Macroradical Reactivity Studied by Electron Spin Resonance. Vol. 3, pp. 688–711.

Bucknall, C. B.: Fracture and Failure of Multiphase Polymers and Polymer Composites. Vol. 27, pp. 121–148.

Burchard, W.: Static and Dynamic Light Scattering from Branched Polymers and Biopolymers. Vol. 48, pp. 1–124.

Bywater, S.: Polymerization Initiated by Lithium and Its Compounds. Vol. 4, pp. 66–110.

Bywater, S.: Preparation and Properties of Star-branched Polymers. Vol. 30, pp. 89–116.

Candau, S., Bastide, J. and *Delsanti, M.:* Structural. Elastic and Dynamic Properties of Swollen Polymer Networks. Vol. 44, pp. 27–72.

Carrick, W. L.: The Mechanism of Olefin Polymerization by Ziegler-Natta Catalysts. Vol. 12, pp. 65–86.

Casale, A. and *Porter, R. S.:* Mechanical Synthesis of Block and Graft Copolymers. Vol. 17, pp. 1–71.

Cerf, R.: La dynamique des solutions de macromolecules dans un champ de vitesses. Vol. 1, pp. 382–450.

Cesca, S., Priola, A. and *Bruzzone, M.:* Synthesis and Modification of Polymers Containing a System of Conjugated Double Bonds. Vol. 32, pp. 1–67.

Chiellini, E., Solaro R., Galli, G. and *Ledwith, A.:* Pptically Active Synthetic Polymers Containing Pendant Carbazolyl Groups. Vol. 62, pp. 143–170.

Cicchetti, O.: Mechanisms of Oxidative Photodegradation and of UV Stabilization of Polyolefins. Vol. 7, pp. 70–112.

Clark, D. T.: ESCA Applied to Polymers. Vol. 24, pp. 125–188.

Coleman, Jr., L. E. and *Meinhardt, N. A.:* Polymerization Reactions of Vinyl Ketones. Vol. 1, pp. 159–179.

Comper, W. D. and *Preston, B. N.:* Rapid Polymer Transport in Concentrated Solutions. Vol. 55, pp. 105–152.

Corner, T.: Free Radical Polymerization — The Synthesis of Graft Copolymers. Vol. 62, pp. 95–142.

Crescenzi, V.: Some Recent Studies of Polyelectrolyte Solutions. Vol. 5, pp. 358–386.

Crivello, J. V.: Cationic Polymerization — Iodonium and Sulfonium Salt Photoinitiators, Vol. 62, pp. 1–48.

Davydov, B. E. and *Krentsel, B. A.:* Progress in the Chemistry of Polyconjugated Systems. Vol. 25, pp. 1–46.

Dettenmaier, M.: Intrinsic Crazes in Polycarbonate Phenomenology and Molecular Interpretation of a New Phenomenon. Vol. 52/53, pp. 57–104

Dobb, M. G. and *McIntyre, J. E.:* Properties and Applications of Liquid-Crystalline Main-Chain Polymers. Vol. 60/61, pp. 61–98.

Döll, W.: Optical Interference Measurements and Fracture Mechanics Analysis of Crack Tip Craze Zones. Vol. 52/53, pp. 105–168

Doi, Y. see *Keii, T.:* Vol. 73/74, pp. 201–248.

Dole, M.: Calorimetric Studies of States and Transitions in Solid High Polymers. Vol. 2, pp. 221–274.

Donnet, J. B., Vidal, A.: Carbon Black-Surface Properties and Interactions with Elastomers. Vol. 76, pp. 103–128.

Dorn, K., Hupfer, B., and *Ringsdorf, H.:* Polymeric Monolayers and Liposomes as Models for Biomembranes How to Bridge the Gap Between Polymer Science and Membrane Biology? Vol. 64, pp. 1–54.

Dreyfuss, P. and *Dreyfuss, M. P.:* Polytetrahydrofuran. Vol. 4, pp. 528–590.

Drobník, J. and *Rypáček, F.:* Soluble Synthetic Polymers in Biological Systems. Vol. 57, pp. 1–50.
Dröscher, M.: Solid State Extrusion of Semicrystalline Copolymers. Vol. 47, pp. 120–138.
Drzal, L. T.: The Interphase in Epoxy Composites. Vol. 75, pp. 1–32.
Dušek, K. and *Prins, W.:* Structure and Elasticity of Non-Crystalline Polymer Networks. Vol. 6, pp. 1–102.
Duncan, R. and *Kopeček, J.:* Soluble Synthetic Polymers as Potential Drug Carriers. Vol. 57, pp. 51–101.

Eastham, A. M.: Some Aspects of the Polymerization of Cyclic Ethers. Vol. 2, pp. 18–50.
Ehrlich, P. and *Mortimer, G. A.:* Fundamentals of the Free-Radical Polymerization of Ethylene. Vol. 7, pp. 386–448.
Eisenberg, A.: Ionic Forces in Polymers. Vol. 5, pp. 59–112.
Eiss, N. S. Jr. see Yorkgitis, E. M. Vol. 72, pp. 79–110.
Elias, H.-G., Bareiss, R. und *Watterson, J. G.:* Mittelwerte des Molekulargewichts und anderer Eigenschaften. Vol. 11, pp. 111–204.
Elsner, G., Riekel, Ch. and *Zachmann, H. G.:* Synchrotron Radiation Physics. Vol. 67, pp. 1–58.
Elyashevich, G. K.: Thermodynamics and Kinetics of Orientational Crystallization of Flexible-Chain Polymers. Vol. 43, pp. 207–246.
Enkelmann, V.: Structural Aspects of the Topochemical Polymerization of Diacetylenes. Vol. 63, pp. 91–136.
Entelis, S. G., Evreinov, V. V., Gorshkov, A. V.: Functionally and Molecular Weight Distribution of Telchelic Polymers. Vol. 76, pp. 129–175.
Evreinov, V. V. see Entelis S. G. Vol. 76, pp. 129–175.

Ferruti, P. and *Barbucci, R.:* Linear Amino Polymers: Synthesis, Protonation and Complex Formation. Vol. 58, pp. 55–92
Finkelmann, H. and *Rehage, G.:* Liquid Crystal Side-Chain Polymers. Vol. 60/61, pp. 99–172.
Fischer, H.: Freie Radikale während der Polymerisation, nachgewiesen und identifiziert durch Elektronenspinresonanz. Vol. 5, pp. 463–530.
Flory, P. J.: Molecular Theory of Liquid Crystals. Vol. 59, pp. 1–36.
Ford, W. T. and *Tomoi, M.:* Polymer-Supported Phase Transfer Catalysts Reaction Mechanisms. Vol. 55, pp. 49–104.
Fradet, A. and *Maréchal, E.:* Kinetics and Mechanisms of Polyesterifications. I. Reactions of Diols with Diacids. Vol. 43, pp. 51–144.
Franz, G.: Polysaccharides in Pharmacy. Vol. 76, pp. 1–30.
Friedrich, K.: Crazes and Shear Bands in Semi-Crystalline Thermoplastics. Vol. 52/53, pp. 225–274.
Fujita, H.: Diffusion in Polymer-Diluent Systems. Vol. 3, pp. 1–47.
Funke, W.: Über die Strukturaufklärung vernetzter Makromoleküle, insbesondere vernetzter Polyesterharze, mit chemischen Methoden. Vol. 4, pp. 157–235.

Gal'braikh, L. S. and *Rigovin, Z. A.:* Chemical Transformation of Cellulose. Vol. 14, pp. 87–130.
Galli, G. see Chiellini, E. Vol. 62, pp. 143–170.
Gallot, B. R. M.: **Preparation and Study of Block Copolymers with Ordered Structures**, Vol. 29, pp. 85–156.
Gandini, A.: The Behaviour of Furan Derivatives in Polymerization Reactions. Vol. 25, pp. 47–96.
Gandini, A. and *Cheradame, H.:* Cationic Polymerization. Initiation with Alkenyl Monomers. Vol. 34/35, pp. 1–289.
Geckeler, K., Pillai, V. N. R., and *Mutter, M.:* Applications of Soluble Polymeric Supports. Vol. 39, pp. 65–94.
Gerrens, H.: Kinetik der Emulsionspolymerisation. Vol. 1, pp. 234–328.
Ghiggino, K. P., Roberts, A. J. and *Phillips, D.:* Time-Resolved Fluorescence Techniques in Polymer and Biopolymer Studies. Vol. 40, pp. 69–167.
Godovsky, Y. K.: Thermomechanics of Polymers. Vol. 76, pp. 31–102.
Goethals, E. J.: The Formation of Cyclic Oligomers in the Cationic Polymerization of Heterocycles. Vol. 23, pp. 103–130.
Gorshkov, A. V. see Entelis, S. G. Vol. 76, pp. 129–175.

Graessley, W. W.: The Etanglement Concept in Polymer Rheology. Vol. 16, pp. 1–179.
Graessley, W. W.: Entagled Linear, Branched and Network Polymer Systems. Molecular Theories. Vol. 47, pp. 67–117.
Grebowicz, J. see Wunderlich, B. Vol. 60/61. pp. 1–60.
Greschner, G. S.: Phase Distribution Chromatography. Possibilities and Limitations. Vol. 73/74, pp. 1–62.

Hagihara, N., Sonogashira, K. and *Takahashi, S.:* Linear Polymers Containing Transition Metals in the Main Chain. Vol. 41, pp. 149–179.
Hasegawa, M.: Four-Center Photopolymerization in the Crystalline State. Vol. 42, pp. 1–49.
Hatano, M.: Induced Circular Dichroism in Biopolymer-Dye System. Vol. 77, pp. 1–121.
Hay, A. S.: Aromatic Polyethers. Vol. 4, pp. 496–527.
Hayakawa, R. and *Wada, Y.:* Piezoelectricity and Related Properties of Polymer Films. Vol. 11, pp. 1–55.
Heidemann, E. and *Roth, W.:* Synthesis and Investigation of Collagen Model Peptides. Vol. 43, pp. 145–205.
Heitz, W.: Polymeric Reagents. Polymer Design, Scope, and Limitations. Vol. 23, pp. 1–23.
Helfferich, F.: Ionenaustausch. Vol. 1, pp. 329–381.
Hendra, P. J.: Laser-Raman Spectra of Polymers. Vol. 6, pp. 151–169.
Hendrix, J.: Position Sensitive "X-ray Detectors". Vol. 67, pp. 59–98.
Henrici-Olivé, G. und *Olivé, S.:* Kettenübertragung bei der radikalischen Polymerisation. Vol. 2, pp. 496–577.
Henrici-Olivé, G. und *Olivé, S.:* Koordinative Polymerisation an löslichen Übergangsmetall-Katalysatoren. Vol. 6, pp. 421–472.
Henrici-Olivé, G. and *Olivé, S.:* Oligomerization of Ethylene with Soluble Transition-Metal Catalysts. Vol. 15, pp. 1–30.
Henrici-Olivé, G. and *Olivé, S.:* Molecular Interactions and Macroscopic Properties of Polyacrylonitrile and Model Substances. Vol. 32, pp. 123–152.
Henrici-Olivé, G. and *Olivé, S.:* The Chemistry of Carbon Fiber Formation from Polyacrylonitrile. Vol. 51, pp. 1–60.
Hermans, Jr., J., Lohr, D. and *Ferro, D.:* Treatment of the Folding and Unfolding of Protein Molecules in Solution According to a Lattic Model. Vol. 9, pp. 229–283.
Higashimura, T. and *Sawamoto, M.:* Living Polymerization and Selective Dimerization: Two Extremes of the Polymer Synthesis by Cationic Polymerization. Vol. 62, pp. 49–94.
Hoffman, A. S.: Ionizing Radiation and Gas Plasma (or Glow) Discharge Treatments for Preparation of Novel Polymeric Biomaterials. Vol. 57, pp. 141–157.
Holzmüller, W.: Molecular Mobility, Deformation and Relaxation Processes in Polymers. Vol. 26, pp. 1–62.
Hutchison, J. and *Ledwith, A.:* Photoinitiation of Vinyl Polymerization by Aromatic Carbonyl Compounds. Vol. 14, pp. 49–86.

Iizuka, E.: Properties of Liquid Crystals of Polypeptides: with Stress on the Electromagnetic Orientation. Vol. 20, pp. 79–107.
Ikada, Y.: Characterization of Graft Copolymers. Vol. 29, pp. 47–84.
Ikada, Y.: Blood-Compatible Polymers. Vol. 57, pp. 103–140.
Imanishi, Y.: Synthese, Conformation, and Reactions of Cyclic Peptides. Vol. 20, pp. 1–77.
Inagaki, H.: Polymer Separation and Characterization by Thin-Layer Chromatography. Vol. 24, pp. 189–237.
Inoue, S.: Asymmetric Reactions of Synthetic Polypeptides. Vol. 21, pp. 77–106.
Ise, N.: Polymerizations under an Electric Field. Vol. 6, pp. 347–376.
Ise, N.: The Mean Activity Coefficient of Polyelectrolytes in Aqueous Solutions and Its Related Properties. Vol. 7, pp. 536–593.
Isihara, A.: Intramolecular Statistics of a Flexible Chain Molecule. Vol. 7, pp. 449–476.
Isihara, A.: Irreversible Processes in Solutions of Chain Polymers. Vol. 5, pp. 531–567.
Isihara, A. and *Guth, E.:* Theory of Dilute Macromolecular Solutions. Vol. 5, pp. 233–260.

Iwatsuki, S.: Polymerization of Quinodimethane Compounds. Vol. 58, pp. 93–120.

Janeschitz-Kriegl, H.: Flow Birefrigence of Elastico-Viscous Polymer Systems. Vol. 6, pp. 170–318.
Jenkins, R. and *Porter, R. S.:* Upertubed Dimensions of Stereoregular Polymers. Vol. 36, pp. 1–20.
Jenngins, B. R.: Electro-Optic Methods for Characterizing Macromolecules in Dilute Solution. Vol. 22, pp. 61–81.
Johnston, D. S.: Macrozwitterion Polymerization. Vol. 42, pp. 51–106.

Kamachi, M.: Influence of Solvent on Free Radical Polymerization of Vinyl Compounds. Vol. 38, pp. 55–87.
Kaneko, M. and *Yamada, A.:* Solar Energy Conversion by Functional Polymers. Vol. 55, pp. 1–48.
Kawabata, S. and *Kawai, H.:* Strain Energy Density Functions of Rubber Vulcanizates from Biaxial Extension. Vol. 24, pp. 89–124.
Keii, T., Doi, Y.: Synthesis of "Living" Polyolefins with Soluble Ziegler-Natta Catalysts and Application to Block Copolymerization. Vol. 73/74, pp. 201–248.
Kennedy, J. P. and *Chou, T.:* Poly(isobutylene-co-β-Pinene): A New Sulfur Vulcanizable, Ozone Resistant Elastomer by Cationic Isomerization Copolymerization. Vol. 21, pp. 1–39.
Kennedy, J. P. and *Delvaux, J. M.:* Synthesis, Characterization and Morphology of Poly(butadiene-g-Styrene). Vol. 38, pp. 141–163.
Kennedy, J. P. and *Gillham, J. K.:* Cationic Polymerization of Olefins with Alkylaluminium Initiators. Vol. 10, pp. 1–33.
Kennedy, J. P. and *Johnston, J. E.:* The Cationic Isomerization Polymerization of 3-Methyl-1-butene and 4-Methyl-1-pentene. Vol. 19, pp. 57–95.
Kennedy, J. P. and *Langer, Jr., A. W.:* Recent Advances in Cationic Polymerization. Vol. 3, pp. 508–580.
Kennedy, J. P. and *Otsu, T.:* Polymerization with Isomerization of Monomer Preceding Propagation. Vol. 7, pp. 369–385.
Kennedy, J. P. and *Rengachary, S.:* Correlation Between Cationic Model and Polymerization Reactions of Olefins. Vol. 14, pp. 1–48.
Kennedy, J. P. and *Trivedi, P. D.:* Cationic Olefin Polymerization Using Alkyl Halide — Alkylaluminium Initiator Systems. I. Reactivity Studies. II. Molecular Weight Studies. Vol. 28, pp. 83–151.
Kennedy, J. P., Chang, V. S. C. and *Guyot, A.:* Carbocationic Synthesis and Characterization of Polyolefins with Si–H and Si–Cl Head Groups. Vol. 43, pp. 1–50.
Khoklov, A. R. and *Grosberg, A. Yu.:* Statistical Theory of Polymeric Lyotropic Liquid Crystals. Vol. 41, pp. 53–97.
Kinloch, A. J.: Mechanics and Mechanisms of Fracture of Thermosetting Epoxy Polymers. Vol. 72, pp. 45–68.
Kissin, Yu. V.: Structures of Copolymers of High Olefins. Vol. 15, pp. 91–155.
Kitagawa, T. and *Miyazawa, T.:* Neutron Scattering and Normal Vibrations of Polymers. Vol. 9, pp. 335–414.
Kitamaru, R. and *Horii, F.:* NMR Approach to the Phase Structure of Linear Polyethylene. Vol. 26, pp. 139–180.
Knappe, W.: Wärmeleitung in Polymeren. Vol. 7, pp. 477–535.
Koenik, J. L. see *Mertzel, E.* Vol. 75, pp. 73–112.
Koenig, J. L.: Fourier Transforms Infrared Spectroscopy of Polymers, Vol. 54, pp. 87–154.
Kolařik, J.: Secondary Relaxations in Glassy Polymers: Hydrophilic Polymethacrylates and Polyacrylates: Vol. 46, pp. 119–161.
Koningsveld, R.: Preparative and Analytical Aspects of Polymer Fractionation. Vol. 7.
Kovacs, A. J.: Transition vitreuse dans les polymers amorphes. Etude phénoménologique. Vol. 3, pp. 394–507.
Krässig, H. A.: Graft Co-Polymerization of Cellulose and Its Derivatives. Vol. 4, pp. 111–156.
Kramer, E. J.: Microscopic and Molecular Fundamentals of Crazing. Vol. 52/53, pp. 1–56
Kraus, G.: Reinforcement of Elastomers by Carbon Black. Vol. 8, pp. 155–237.
Kreutz, W. and *Welte, W.:* A General Theory for the Evaluation of X-Ray Diagrams of Biomembranes and Other Lamellar Systems. Vol. 30, pp. 161–225.

Krimm, S.: Infrared Spectra of High Polymers. Vol. 2, pp. 51–72.
Kuhn, W., Ramel, A., Walters, D. H., Ebner, G. and *Kuhn, H. J.:* The Production of Mechanical Energy from Different Forms of Chemical Energy with Homogeneous and Cross-Striated High Polymer Systems. Vol. 1, pp. 540–592.
Kunitake, T. and *Okahata, Y.:* Catalytic Hydrolysis by Synthetic Polymers. Vol. 20, pp. 159–221.
Kurata, M. and *Stockmayer, W. H.:* Intrinsic Viscosities and Unperturbed Dimensions of Long Chain Molecules. Vol. 3, pp. 196–312.

Ledwith, A. and *Sherrington, D. C.:* Stable Organic Cation Salts: Ion Pair Equilibria and Use in Cationic Polymerization. Vol. 19, pp. 1–56.
Ledwith, A. see Chiellini, E. Vol. 62, pp. 143–170.
Lee, C.-D. S. and *Daly, W. H.:* Mercaptan-Containing Polymers. Vol. 15, pp. 61–90.
Lindberg, J. J. and *Hortling, B.:* Cross Polarization — Magic Angle Spinning NMR Studies of Carbohydrates and Aromatic Polymers. Vol. 66, pp. 1–22.
Lipatov, Y. S.: Relaxation and Viscoelastic Properties of Heterogeneous Polymeric Compositions. Vol. 22, pp. 1–59.
Lipatov, Y. S.: The Iso-Free-Volume State and Glass Transitions in Amorphous Polymers: New Development of the Theory. Vol. 26, pp. 63–104.
Lustoň, J. and *Vašš, F.:* Anionic Copolymerization of Cyclic Ethers with Cyclic Anhydrides. Vol. 56, pp. 91–133.

Madec, J.-P. and *Maréchal, E.:* Kinetics and Mechanisms of Polyesterifications. II. Reactions of Diacids with Diepoxides. Vol. 71, pp. 153–228.
Mano, E. B. and *Coutinho, F. M. B.:* Grafting on Polyamides. Vol. 19, pp. 97–116.
Maréchal, E. see Madec, J.-P. Vol. 71, pp. 153–228.
Mark, J. E.: The Use of Model Polymer Networks to Elucidate Molecular Aspects of Rubberlike Elasticity. Vol. 44, pp. 1–26.
Mark, J. E. see Queslel, J. P. Vol. 71, pp. 229–248.
Maser, F., Bode, K., Pillai, V. N. R. and *Mutter, M.:* Conformational Studies on Model Peptides. Their Contribution to Synthetic, Structural and Functional Innovations on Proteins. Vol. 65, pp. 177–214.
McGrath, J. E. see Yorkgitis, E. M. Vol. 72, pp. 79–110.
McIntyre, J. E. see Dobb, M. G. Vol. 60/61, pp. 61–98.
Meerwall v., E., D.: Self-Diffusion in Polymer Systems. Measured with Field-Gradient Spin Echo NMR Methods, Vol. 54, pp. 1–29.
Mengoli, G.: Feasibility of Polymer Film Coating Through Electroinitiated Polymerization in Aqueous Medium. Vol. 33, pp. 1–31.
Mertzel, E., Koenik, J. L. Β Application of FT-IR and NMR to Epoxy Resins. Vol. 75, pp. 73–112.
Meyerhoff, G.: Die viscosimetrische Molekulargewichtsbestimmung von Polymeren. Vol. 3, pp. 59–105.
Millich, F.: Rigid Rods and the Characterization of Polyisocyanides. Vol. 19, pp. 117–141.
Möller, M.: Cross Polarization — Magic Angle Sample Spinning NMR Studies. With Respect to the Rotational Isomeric States of Saturated Chain Molecules. Vol. 66, pp. 59–80.
Morawetz, H.: Specific Ion Binding by Polyelectrolytes. Vol. 1, pp. 1–34.
Morgan, R. J.: Structure-Property Relations of Epoxies Used as Composite Matrices. Vol. 72, pp. 1–44.
Morin, B. P., Breusova, I. P. and *Rogovin, Z. A.:* Structural and Chemical Modifications of Cellulose by Graft Copolymerization. Vol. 42, pp. 139–166.
Mulvaney, J. E., Oversberger, C. C. and *Schiller, A. M.:* Anionic Polymerization. Vol. 3, pp. 106–138.

Nakase, Y., Kurijama, I. and *Odajima, A.:* Analysis of the Fine Structure of Poly(Oxymethylene) Prepared by Radiation-Induced Polymerization in the Solid State. Vol. 65, pp. 79–134.
Neuse, E.: **Aromatic Polybenzimidazoles. Syntheses, Properties, and Applications.** Vol. 47, pp. 1–42.
Nicolais, L. see Apicella, A. Vol. 72, pp. 69–78.
Nuyken, O., Weidner, R.: Graft and Block Copolymers via Polymeric Azo Initiators. Vol. 73/74, pp. 145–200.

Ober, Ch. K., Jin, J.-I. and *Lenz, R. W.:* Liquid Crystal Polymers with Flexible Spacers in the Main Chain. Vol. 59, pp. 103–146.
Okubo, T. and *Ise, N.:* Synthetic Polyelectrolytes as Models of Nucleic Acids and Esterases. Vol. 25, pp. 135–181.
Osaki, K.: Viscoelastic Properties of Dilute Polymer Solutions. Vol. 12, pp. 1–64.
Oster, G. and *Nishijima, Y.:* Fluorescence Methods in Polymer Science. Vol. 3, pp. 313–331.
Otsu, T. see Sato, T. Vol. 71, pp. 41–78.
Overberger, C. G. and *Moore, J. A.:* Ladder Polymers. Vol. 7, pp. 113–150.

Packirisamy, S. see Biswas, M. Vol. 70, pp. 71–118.
Papkov, S. P.: Liquid Crystalline Order in Solutions of Rigid-Chain Polymers. Vol. 59, pp. 75–102.
Patat, F., Killmann, E. und *Schiebener, C.:* Die Absorption von Makromolekülen aus Lösung. Vol. 3, pp. 332–393.
Patterson, G. D.: Photon Correlation Spectroscopy of Bulk Polymers. Vol. 48, pp. 125–159.
Penczek, S., Kubisa, P. and *Matyjaszewski, K.:* Cationic Ring-Opening Polymerization of Heterocyclic Monomers. Vol. 37, pp. 1–149.
Penczek, S., Kubisa, P. and *Matyjaszewski, K.:* Cationic Ring-Opening Polymerization; 2. Synthetic Applications. Vol. 68/69, pp. 1–298.
Peticolas, W. L.: Inelastic Laser Light Scattering from Biological and Synthetic Polymers. Vol. 9, pp. 285–333.
Petropoulos, J. H.: Membranes with Non-Homogeneous Sorption Properties. Vol. 64, pp. 85–134.
Pino, P.: Optically Active Addition Polymers. Vol. 4, pp. 393–456.
Pitha, J.: Physiological Activities of Synthetic Analogs of Polynucleotides. Vol. 50, pp. 1–16.
Platé, N. A. and *Noak, O. V.:* A Theoretical Consideration of the Kinetics and Statistics of Reactions of Functional Groups of Macromolecules. Vol. 31, pp. 133–173.
Platé, N. A. see Shibaev, V. P. Vol. 60/61, pp. 173–252.
Plesch, P. H.: The Propagation Rate-Constants in Cationic Polymerisations. Vol. 8, pp. 137–154.
Porod, G.: Anwendung und Ergebnisse der Röntgenkleinwinkelstreuung in festen Hochpolymeren. Vol. 2, pp. 363–400.
Pospíšil, J.: Transformations of Phenolic Antioxidants and the Role of Their Products in the Long-Term Properties of Polyolefins. Vol. 36, pp. 69–133.
Postelnek, W., Coleman, L. E., and *Lovelace, A. M.:* Fluorine-Containing Polymers. I. Fluorinated Vinyl Polymers with Functional Groups, Condensation Polymers, and Styrene Polymers. Vol. 1, pp. 75–113.

Queslel, J. P. and *Mark, J. E.:* Molecular Interpretation of the Moduli of Elastomeric Polymer Networks of Know Structure. Vol. 65, pp. 135–176.
Queslel, J. P. and *Mark, J. E.:* Swelling Equilibrium Studies of Elastomeric Network Structures. Vol. 71, pp. 229–248.

Rehage, G. see Finkelmann, H. Vol. 60/61, pp. 99–172.
Rempp, P. F. and *Franta, E.:* Macromonomers: Synthesis, Characterization and Applications. Vol. 58, pp. 1–54.
Rempp, P., Herz, J., and *Borchard, W.:* Model Networks. Vol. 26, pp. 107–137.
Richards, R. W.: Small Angle Neutron Scattering from Block Copolymers. Vol. 71, pp. 1–40.
Rigbi, Z.: Reinforcement of Rubber by Carbon Black. Vol. 36, pp. 21–68.
Rogovin, Z. A. and *Gabrielyan, G. A.:* Chemical Modifications of Fibre Forming Polymers and Copolymers of Acrylonitrile. Vol. 25, pp. 97–134.
Roha, M.: Ionic Factors in Steric Control. Vol. 4, pp. 353–392.
Roha, M.: The Chemistry of Coordinate Polymerization of Dienes. Vol. 1, pp. 512–539.
Rostami, S. see Walsh, D. J. Vol. 70, pp. 119–170.
Rozengerk, v. A. B Kinetics, Thermodynamics and Mechanism of Reactions of Epoxy Oligomers with Amines. Vol. 75, pp. 113–166.

Safford, G. J. and *Naumann, A. W.:* Low Frequency Motions in Polymers as Measured by Neutron Inelastic Scattering. Vol. 5, pp. 1–27.
Sato, T. and *Otsu, T.:* Formation of Living Propagating Radicals in Microspheres and Their Use in the Synthesis of Block Copolymers. Vol. 71, pp. 41–78.
Sauer, J. A. and *Chen, C. C.:* Crazing and Fatigue Behavior in One and Two Phase Glassy Polymers. Vol. 52/53, pp. 169–224
Sawamoto, M. see Higashimura, T. Vol. 62, pp. 49–94.
Schmidt, R. G., Bell, J. P.: Epoxy Adhesion to Metals. Vol. 75, pp. 33–72.
Schuerch, C.: The Chemical Synthesis and Properties of Polysaccharides of Biomedical Interest. Vol. 10, pp. 173–194.
Schulz, R. C. und *Kaiser, E.:* Synthese und Eigenschaften von optisch aktiven Polymeren. Vol. 4, pp. 236–315.
Seanor, D. A.: Charge Transfer in Polymers. Vol. 4, pp. 317–352.
Semerak, S. N. and *Frank, C. W.:* Photophysics of Excimer Formation in Aryl Vinyl Polymers, Vol. 54, pp. 31–85.
Seidl, J., Malinský, J., Dušek, K. und *Heitz, W.:* Makroporöse Styrol-Divinylbenzol-Copolymere und ihre Verwendung in der Chromatographie und zur Darstellung von Ionenaustauschern. Vol. 5, pp. 113–213.
Semjonow, V.: Schmelzviskositäten hochpolymerer Stoffe. Vol. 5, pp. 387–450.
Semlyen, J. A.: Ring-Chain Equilibria and the Conformations of Polymer Chains. Vol. 21, pp. 41–75.
Sen, A.: The Copolymerization of Carbon Monoxide with Olefins. Vol. 73/74, pp. 125–144.
Sharkey, W. H.: Polymerizations Through the Carbon-Sulphur Double Bond. Vol. 17, pp. 73–103.
Shibaev, V. P. and *Platé, N. A.:* Thermotropic Liquid-Crystalline Polymers with Mesogenic Side Groups. Vol. 60/61, pp. 173–252.
Shimidzu, T.: Cooperative Actions in the Nucleophile-Containing Polymers. Vol. 23, pp. 55–102.
Shutov, F. A.: Foamed Polymers Based on Reactive Oligomers, Vol. 39, pp. 1–64.
Shutov, F. A.: Foamed Polymers. Cellular Structure and Properties. Vol. 51, pp. 155–218.
Shutov, F. A.: Syntactic Polymer Foams. Vol. 73/74, pp. 63–124.
Siesler, H. W.: Rheo-Optical Fourier-Transform Infrared Spectroscopy: Vibrational Spectra and Mechanical Properties of Polymers. Vol. 65, pp. 1–78.
Silvestri, G., Gambino, S., and *Filardo, G.:* Electrochemical Production of Initiators for Polymerization Processes. Vol. 38, pp. 27–54.
Sixl, H.: Spectroscopy of the Intermediate States of the Solid State Polymerization Reaction in Diacetylene Crystals. Vol. 63, pp. 49–90.
Slichter, W. P.: The Study of High Polymers by Nuclear Magnetic Resonance. Vol. 1, pp. 35–74.
Small, P. A.: Long-Chain Branching in Polymers. Vol. 18.
Smets, G.: Block and Graft Copolymers. Vol. 2, pp. 173–220.
Smets, G.: Photochromic Phenomena in the Solid Phase. Vol. 50, pp. 17–44.
Sohma, J. and *Sakaguchi, M.:* ESR Studies on Polymer Radicals Produced by Mechanical Destruction and Their Reactivity. Vol. 20, pp. 109–158.
Solaro, R. see Chiellini, E. Vol. 62, pp. 143–170.
Sotobayashi, H. und *Springer, J.:* Oligomere in verdünnten Lösungen. Vol. 6, pp. 473–548.
Sperati, C. A. and *Starkweather, Jr., H. W.:* Fluorine-Containing Polymers. II. Polytetrafluoroethylene. Vol. 2, pp. 465–495.
Spiess, H. W.: Deuteron NMR — A new Tool for Studying Chain Mobility and Orientation in Polymers. Vol. 66, pp. 23–58.
Sprung, M. M.: Recent Progress in Silicone Chemistry. I. Hydrolysis of Reactive Silane Intermediates, Vol. 2, pp. 442–464.
Stahl, E. and *Brüderle, V.:* Polymer Analysis by Thermofractography. Vol. 30, pp. 1–88.
Stannett, V. T., Koros, W. J., Paul, D. R., Lonsdale, H. K., and *Baker, R. W.:* Recent Advances in Membrane Science and Technology. Vol. 32, pp. 69–121.
Staverman, A. J.: Properties of Phantom Networks and Real Networks. Vol. 44, pp. 73–102.
Stauffer, D., Coniglio, A. and *Adam, M.:* Gelation and Critical Phenomena. Vol. 44, pp. 103–158.
Stille, J. K.: Diels-Alder Polymerization. Vol. 3, pp. 48–58.
Stolka, M. and *Pai, D.:* Polymers with Photoconductive Properties. Vol. 29, pp. 1–45.
Stuhrmann, H.: Resonance Scattering in Macromolecular Structure Research. Vol. 67, pp. 123–164.
Subramanian, R. V.: Electroinitiated Polymerization on Electrodes. Vol. 33, pp. 35–58.

Sumitomo, H. and *Hashimoto, K.:* Polyamides as Barrier Materials. Vol. 64, pp. 55–84.
Sumitomo, H. and *Okada, M.:* Ring-Opening Polymerization of Bicyclic Acetals, Oxalactone, and Oxalactam. Vol. 28, pp. 47–82.
Szegö, L.: Modified Polyethylene Terephthalate Fibers. Vol. 31, pp. 89–131.
Szwarc, M.: Termination of Anionic Polymerization. Vol. 2, pp. 275–306.
Szwarc, M.: The Kinetics and Mechanism of N-carboxy-α-amino-acid Anhydride (NCA) Polymerization to Poly-amino Acids. Vol. 4, pp. 1–65.
Szwarc, M.: Thermodynamics of Polymerization with Special Emphasis on Living Polymers. Vol. 4, pp. 457–495.
Szwarc, M.: Living Polymers and Mechanisms of Anionic Polymerization. Vol. 49, pp. 1–175.

Takahashi, A. and *Kawaguchi, M.:* The Structure of Macromolecules Adsorbed on Interfaces. Vol. 46, pp. 1–65.
Takemoto, K. and *Inaki, Y.:* Synthetic Nucleic Acid Analogs. Preparation and Interactions. Vol. 41, pp. 1–51.
Tani, H.: Stereospecific Polymerization of Aldehydes and Epoxides. Vol. 11, pp. 57–110.
Tate, B. E.: Polymerization of Itaconic Acid and Derivatives. Vol. 5, pp. 214–232.
Tazuke, S.: Photosensitized Charge Transfer Polymerization. Vol. 6, pp. 321–346.
Teramoto, A. and *Fujita, H.:* Conformation-dependent Properties of Synthetic Polypeptides in the Helix-Coil Transition Region. Vol. 18, pp. 65–149.
Theocaris, P. S.: The Mesophase and its Influence on the Mechanical Behavior of Composites. Vol. 66, pp. 149–188.
Thomas, W. M.: Mechanismus of Acrylonitrile Polymerization. Vol. 2, pp. 401–441.
Tieke, B.: Polymerization of Butadiene and Butadiyne (Diacetylene) Derivatives in Layer Structures. Vol. 71, pp. 79–152.
Tobolsky, A. V. and *DuPré, D. B.:* Macromolecular Relaxation in the Damped Torsional Oscillator and Statistical Segment Models. Vol. 6, pp. 103–127.
Tosi, C. and *Ciampelli, F.:* Applications of Infrared Spectroscopy to Ethylene-Propylene Copolymers. Vol. 12, pp. 87–130.
Tosi, C.: Sequence Distribution in Copolymers: Numerical Tables. Vol. 5, pp. 451–462.
Tran, C. see Yorkgitis, E. M. Vol. 72, pp. 79–110.
Tsuchida, E. and *Nishide, H.:* Polymer-Metal Complexes and Their Catalytic Activity. Vol. 24, pp. 1–87.
Tsuji, K.: ESR Study of Photodegradation of Polymers. Vol. 12, pp. 131–190.
Tsvetkov, V. and *Andreeva, L.:* Flow and Electric Birefringence in Rigid-Chain Polymer Solutions. Vol. 39, pp. 95–207.
Tuzar, Z., Kratochvil, P., and *Bohdanecký, M.:* Dilute Solution Properties of Aliphatic Polyamides. Vol. 30, pp. 117–159.

Uematsu, I. and *Uematsu, Y.:* Polypeptide Liquid Crystals. Vol. 59, pp. 37–74.

Valvassori, A. and *Sartori, G.:* Present Status of the Multicomponent Copolymerization Theory. Vol. 5, pp. 28–58.
Vidal, A. see Donnet, J. B. Vol. 76, pp. 103–128.
Viovy, J. L. and *Monnerie, L.:* Fluorescence Anisotropy Technique Using Synchrotron Radiation as a Powerful Means for Studying the Orientation Correlation Functions of Polymer Chains. Vol. 67, pp. 99–122.
Voigt-Martin, I.: Use of Transmission Electron Microscopy to Obtain Quantitative Information About Polymers. Vol. 67, pp. 195–218.
Voorn, M. J.: Phase Separation in Polymer Solutions. Vol. 1, pp. 192–233.

Walsh, D. J., Rostami, S.: The Miscibility of High Polymers: The Role of Specific Interactions. Vol. 70, pp. 119–170.

Ward, I. M.: Determination of Molecular Orientation by Spectroscopic Techniques. Vol. 66, pp. 81–116.

Ward, I.M.: The Preparation, Structure and Properties of Ultra-High Modulus Flexible Polymers. Vol. 70, pp. 1–70.

Weidner, R. see *Nuyken, O.:* Vol. 73/74, pp. 145–200.

Werber, F. X.: Polymerization of Olefins on Supported Catalysts. Vol. 1, pp. 180–191.

Wichterle, O., Šebenda, J., and *Králiček, J.:* The Anionic Polymerization of Caprolactam. Vol. 2, pp. 578–595.

Wilkes, G. L.: The Measurement of Molecular Orientation in Polymeric Solids. Vol. 8, pp. 91–136.

Wilkes, G. L. see Yorkgitis, E. M. Vol. 72, pp. 79–110.

Williams, G.: Molecular Aspects of Multiple Dielectric Relaxation Processes in Solid Polymers. Vol. 33, pp. 59–92.

Williams, J. G.: Applications of Linear Fracture Mechanics. Vol. 27, pp. 67–120.

Wöhrle, D.: Polymere aus Nitrilen. Vol. 10, pp. 35–107.

Wöhrle, D.: Polymer Square Planar Metal Chelates for Science and Industry. Synthesis, Properties and Applications. Vol. 50, pp. 45–134.

Wolf, B. A.: Zur Thermodynamik der enthalpisch und der entropisch bedingten Entmischung von Polymerlösungen. Vol. 10, pp. 109–171.

Woodward, A. E. and *Sauer, J. A.:* The Dynamic Mechanical Properties of High Polymers at Low Temperatures. Vol. 1, pp. 114–158.

Wunderlich, B.: Crystallization During Polymerization. Vol. 5, pp. 568–619.

Wunderlich, B. and *Baur, H.:* Heat Capacities of Linear High Polymers. Vol. 7, pp. 151–368.

Wunderlich, B. and *Grebowicz, J.:* Thermotropic Mesophases and Mesophase Transitions of Linear, Flexible Macromolecules. Vol. 60/61, pp. 1–60.

Wrasidlo, W.: Thermal Analysis of Polymers. Vol. 13, pp. 1–99.

Yamashita, Y.: Random and Black Copolymers by Ring-Opening Polymerization. Vol. 28, pp. 1–46.

Yamazaki, N.: Electrolytically Initiated Polymerization. Vol. 6, pp. 377–400.

Yamazaki, N. and *Higashi, F.:* New Condensation Polymerizations by Means of Phosphorus Compounds. Vol. 38, pp. 1–25.

Yokoyama, Y. and *Hall, H. K.:* Ring-Opening Polymerization of Atom-Bridged and Bond-Bridged Bicyclic Ethers, Acetals and Orthoesters. Vol. 42, pp. 107–138.

Yorkgitis, E. M., Eiss, N. S. Jr., Tran, C., Wilkes, G. L. and *McGrath, J. E.:* Siloxane-Modified Epoxy Resins. Vol. 72, pp. 79–110.

Yoshida, H. and *Hayashi, K.:* Initiation Process of Radiation-induced Ionic Polymerization as Studied by Electron Spin Resonance. Vol. 6, pp. 401–420.

Young, R. N., Quirk, R. P. and *Fetters, L. J.:* Anionic Polymerizations of Non-Polar Monomers Involving Lithium. Vol. 56, pp. 1–90.

Yuki, H. and *Hatada, K.:* Stereospecific Polymerization of Alpha-Substituted Acrylic Acid Esters. Vol. 31, pp. 1–45.

Zachmann, H. G.: Das Kristallisations- und Schmelzverhalten hochpolymerer Stoffe. Vol. 3, pp. 581–687.

Zaikov, G. E. see Aseeva, R. M. Vol. 70, pp. 171–230.

Zakharov, V. A., Bukatov, G. D., and *Yermakov, Y. I.:* On the Mechanism of Olifin Polymerization by Ziegler-Natta Catalysts. Vol. 51, pp. 61–100.

Zambelli, A. and *Tosi, C.:* Stereochemistry of Propylene Polymerization. Vol. 15, pp. 31–60.

Zucchini, U. and *Cecchin, G.:* Control of Molecular-Weight Distribution in Polyolefins Synthesized with Ziegler-Natta Catalytic Systems. Vol. 51, pp. 101–154.

Subject Index

Absorption flattering 100
2-Acetamido-2-deoxy-D-glucosamine 92
N-Acetyl-L-l-naphthylalanine ethyl ester 70
Acridine orange (AO) 72
N-Acylamino acids 97
Adriamycin 46
Aggregates of chiral amphiphiles 99f.
Alkaline phosphatase 88
Amphiphiles 99
Anisotory factor 67
Anomeric configuration in saccharides 93f.
Aplysia myoglobin 54
Apomyoglobin-chlorophyll complexes 80
Aromatic compounds bound to proteins 72f.
Asymmetric carbon 2
— molecules 12
Asymmetrically perturbed field mechanism 21f.
Axial symmetry 12

Bacteriochlorophyll-protein complexes 82, 99
Bacteriochlorophyll-protein, light-harvesting 86
Benzene 110
Benzoates exciton rule 93f.
Bis(acetylacetonato)copper(II) 28
Bis(phenanthroline)dichlororuthenium(II) 48

Ca^{2+}-binding protein 62
Calmodulin 63, 77, 87
Calycanthine 15
(+)-Camphor-10-sulfonic acid 103
Carboxymethylcellulose 90
CD (circular dichroism) 1, 7ff.
—, calibration 103
— measurements 103
—, pitch-band 97f.
—, pitch-dependent 94
—, split 93
Charge displacement 17
— distribution 20
—-transfer-induced circular dichroism (CTICD) 22, 29f.
Chitan 92
Chitin 91
Chiral amphiphiles, aggregates of 99f.
— smectic C phases 98

Chirality 1, 9, 12
— rule 93
Chlorophyll-apomyoglobin complexes 80
— bound to proteins 78f.
— dimer 83
Cholesteric liquid crystalline phase 68
— phases 94f.
— — pitch 95
— pitch 34
Chou and Fasman method 54
Christiansen effect 97
Chromatic emulsions 97
— suspensions 97
Chromatium vinosum 83
Chromophores, side-chain 66
Circular dichroism (CD) 1, 7ff.
Circularly birefringent 4
— polarized light 35
CNDO/S-CI calculation 37
CNDO/S-CI method 113
CNDO/S-MO method 21
Collapsed monolayer 101
Conformational change of proteins 61f.
Coupled oscillator interaction 79
— — mechanism 12ff.
CTICD 22, 29f.
CT transition 29
Cu(II)-β-diketonates 27
β-Cyclodextrin 37
— complex 113
Cytochrome P-450 60, 77

Davydov pair of CD 68
—-type splitting pair 70
Dextrorotatory 6
D-homo-5α-androstane-3β,15β-bis(p-dimethylaminobenzoate) 71
Dialuminium stearate 97
Dibenzoate exciton rule 93
DICD 22, 24f.
(+)-Diethyltartrate 25
Dipole-dipole interaction 20
—-length procedure 20
—-velocity procedure 20

Dispersion force-induced circular dichroism (DICD) 24f.
Dissymmetric molecules 2
Dissymmetry 9
DNA 40ff.
Drude formula, one-term 7

Electron donor-acceptor complex 29
Elliptically polarized light 7
Ellipticity 7
—, molar 9
—, specific 8
Emulsion, chromatic 97
Enantiomers 1
Exciton coupling 71, 77
Experimental considerations 102ff.
Extended Hückel method 21

Faraday A term, B term 107
Faraday C term 108
Faraday effect 6, 107
^{19}F NMR 77

Gaussian fitting method 77
β-D-Glucan 90
α-1,4-D-Glucan (amylose) 90
α-Glycol dibenzoates 19

Handness 1
— in liquid crystals 6
HBICD 22, 26f.
Helical sense in cholesteric phases 94f.
Helicity 34
Helix 52
—, right-handed 4, 35
Hemoglobin 78, 82
Hexafluoro-2-propanol (HFIP) 91
HIICD 22, 37f.
Hückel method, extended 21
Hydrogen bonding-induced circular dichroism (HBICD) 26f.
Hydropathic index 60
Hydrophobic interaction-induced circular dichroism (HIICD) 37f.
12-Hydroxyoctadecanoic acid 97

ICD 1, 22, 24
—, first-order 24
— in liquid crystalline phases 94
—, second-order term 24
ICICD 22, 27f.
ICM 83
Improper axis of symmetry 11
Induced circular dichroism (ICD) 1, 12
— optical activity 22f.
Intracytoplasmic membrane (ICM) 83

Ionic coupling-induced circular dichroism (ICICD) 27

Kirkwood's theory 78
Kronig-Kramers integral transform 9
Kyte and Doolittle method 60, 86

α-Lactalbumin 62
LCICD 22, 34f., 94
LCP 4, 8
Left-circularly polarized light (LCP) 4
LICD 22, 27f.
Ligation-induced circular dichroism (LICD) 27
Linearly polarized wave 4
Liquid crystal-induced circular dichroism (LCICD) 34f.
— crystals 94
Lorentz correction 9
— factor 6

Magnetic circular dichroism (MCD) 30, 87, 107ff.
— moment 10
— rotation (MOR) 107
Maxwell's equation 94
MCD 69, 82, 87, 107
— measurements 103
Membrane-bound proteins 60
— particles 100
Membranes, intracytoplasmic 83
Mesophase 34
Metal binding to proteins 87f.
Methyl orange (MO) 74
Moffitt-Yang equation 7, 51
Molar ellipticity 9
Molecular orbital calculation 69
— polarizability 10
— symmetry 2
Molecules-in-molecule method 20
MOR 107
Myoglobin 54ff., 78, 80

Naphthalenes 69, 113
Nickel(II)tartrate 103
Nicotine 27
Nucleic acid-dye systems 40ff.
— —-protein systems 49f.

Octant rule 21
Octyl glucoside 100
One-electron approximation 21
One-term Drude formula 7
Optical activity 1,2ff.
— —, theory of 9ff.
— —, quantum-mechanical treatment 10
Optical rotation 1, 4ff.

Subject Index

Optical rotatory dispersion (ORD) 1, 6, 51
Osteocalcin 63

Pariser-Parr-Pople method 15
Pauli matrices 95
PBDG 67
PCLA 67
PCLG 68
Phenylethylamine 27
Phenylglycine 67
pH-jump method 62
Phosphatase, alkaline 88
Pitch-band CD 97f.
Pitch-dependent CD 94
Plane-polarized wave 4
PLGA 72
PLL = Poly(L-lysine) 55, 74
PLNA 70
PNVC = Poly(vinyl carbazole) 69
Point groups 13
Polarizability, molecular 10
Polarization 4
— direction 5
Polarized light 3f.
— —, circularly 35
— —, elliptically 7
Polarizers 105
Poly(α,γ-diaminobutyric acid) 77
Poly(α,L-glutamic acid) 72
Poly(β-benzyl-L-aspartate) 67
Poly(γ-benzyl-D-glutamate) 67
Poly(γ-[2-(9-carbazolyl)ethyl]-L-glutamate) 68
Poly(γ-p-nitrobenzyl-L-glutamate) 67
Poly(L-α,γ-diaminobutyric acid) 55
Poly(L-lysine) = PLL 55, 74
Poly(L-naphthylalanine) 70
Poly(L-ornithine) 55
Poly(L-phenylalanine) 70
Poly(L-tryptophan) 70
Poly(L-tyrosine) 70
Poly(S-carboxymethyl-L-cysteine) 76
Poly(vinyl carbazole) = PNVC 69
Polymer aggregate 74
Poly(naphthylmethyl-L-glutamate) 71
Polysaccharide conformation analysis 90f.
Polysaccharide-dye systems 89ff.
Polysaccharide-dye systems 89ff.
Protein-bacteriochlorophyll complexes 82, 99
—- dye systems 50ff.
Proteins, aromatic compounds bound to 72f.
—, chlorophyll bound to 78f.
—, conformational change 61

—, membrane-bound 60
—, metal binding to 87f.
—, prediction of secondary structure 54
$Pt(NH_3)_2Cl_2$ 47

Rapid scanning spectropolarimeters 89
RCP 3, 8
Refractive index 34
Rhodopseudomonas capsulata 86
— *palustris* 83
— *sphaeroides* 86
Rhodospirillum rubrum 86
Right-circularly polarized light (RCP) 3
—-handed helix 4
RNA 40
Rotational-reflection axis 2
$Ru(phen)_2Cl_2$ 48

SCF-MO method 21
Secondary structure 51
— — in proteins, prediction 54
β Sheet 52
Side-chain chromophores 66
—-— — on saccharides 90ff.
Soret CD band 79
Specific ellipticity 8
Stopped-flow system 89
Stress modulators 105
Superimposability 2, 9
Superoxide dismutase 87
Suspensions, chromatic 97
Symmetry, axial 12
— considerations 10f.
— elements 13
—, improper axis of 11
Synchrotron orbital radiation (SOR) 93, 106

TCNE 29, 30
Temperature jump 62
—-— system 89
Tetracyanoethylene (TCNE) 29, 30
Trifluoperazine (TFP) 77
Tris(ethylenediamine)cobalt(III)ions 1
Tween 80 35

Vacuum ultraviolet circular dichroism (VUV CD) 89f., 93, 94

Z-DNA 40ff.

Anionic Polymerizations of Non-Polar Monomers Involving Lithium

R. N. Young, R. P. Quirk and L. J. Fetters
Advances Polymer Science, Vol. 56 (1984)

Errata

Page 25, line 9: "t-butyllithium $^{44,106-108)}$" should be "t-butyllithium $^{44,107,108)}$"

Page 64, Eq. (52): $[DLi]_0^{1/2}$ should be $[DL]_0^{-1/2}$

Page 64, line 12: "ethylene $^{272)}$" should be "ethylene $^{271a)}$".
Ref. 271a — Liata, Z., and Szwarc, M., Macromolecules 2, 412 (1969) — was inadvertently omitted from the list of references.

Page 64, line 29: $[DPE] > [S^-M^+]$ should be $[DPE] < [S^-M^+]$

Page 64, line 30: "cross-association was not considered" should be "cross-association did not have to be considered."

Page 74, line 22: "term" should be "termination"

Page 78, line 22: "forme" should be "formed".